Universitext

Springer
Berlin
Heidelberg
New York
Hong Kong
London
Milan
Paris
Tokyo

Thomas Mikosch

Non-Life Insurance Mathematics

An Introduction with Stochastic Processes

 Springer

Thomas Mikosch

University of Copenhagen
Lab. Actuarial Mathematics
Inst. Mathematical Sciences
Universitetsparken 5
2100 Copenhagen
Denmark
e-mail: mikosch@math.ku.dk

Cataloging-in-Publication Data applied for
A catalog record for this book is available from the Library of Congress.
Bibliographic information published by Die Deutsche Bibliothek
Die Deutsche Bibliothek lists this publication in the Deutsche Nationalbibliografie;
detailed bibliographic data is available in the Internet at http://dnb.ddb.de

Mathematics Subject Classification (2000): 91B30, 60G35, 60K10

ISBN 3-540-40650-6 Springer-Verlag Berlin Heidelberg New York

Springer-Verlag Berlin Heidelberg New York
Springer-Verlag is a part of Springer Science+Business Media

springeronline.com

© Springer-Verlag Berlin Heidelberg 2004
Printed in Germany

Cover design: *design & production* GmbH, Heidelberg
Cover picture: courtesy of the Institut des Hautes Études Scientifiques, Bures-sur-Yvette
Typeset by the author using a Springer LaTeX macro package

Printed on acid-free paper   41/3142db- 5 4 3 2 1 0

Contents

Part II Experience Rating

Preface

To the outside world, insurance mathematics does not appear as a challenging topic. In fact, everyone has to deal with matters of insurance at various times of one's life. Hence this is quite an interesting perception of a field which constitutes one of the bases of modern society. There is no doubt that modern economies and states would not function without institutions which guarantee reimbursement to the individual, the company or the organization for its losses, which may occur due to natural or man-made catastrophes, fires, floods, accidents, riots, etc. The idea of insurance is part of our civilized world. It is based on the mutual trust of the insurer and the insured.

It was realized early on that this mutual trust must be based on science, not on belief and speculation. In the 20th century the necessary tools for dealing with matters of insurance were developed. These consist of probability theory, statistics and stochastic processes. The Swedish mathematicians Filip Lundberg and Harald Cramér were pioneers in these areas. They realized in the first half of the 20th century that the theory of stochastic processes provides the most appropriate framework for modeling the claims arriving in an insurance business. Nowadays, the Cramér-Lundberg model is one of the backbones of non-life insurance mathematics. It has been modified and extended in very different directions and, morever, has motivated research in various other fields of applied probability theory, such as queuing theory, branching processes, renewal theory, reliability, dam and storage models, extreme value theory, and stochastic networks.

The aim of this book is to bring some of the standard stochastic models of non-life insurance mathematics to the attention of a wide audience which, hopefully, will include actuaries and also other applied scientists. The primary objective of this book is to provide the undergraduate actuarial student with an introduction to non-life insurance mathematics. I used parts of this text in the course on basic non-life insurance for 3rd year mathematics students at the Laboratory of Actuarial Mathematics of the University of Copenhagen. But I am convinced that the content of this book will also be of interest to others who have a background on probability theory and stochastic processes and would like to learn about applied stochastic processes. Insurance mathematics is a part of applied probability theory. Moreover, its mathematical tools are also used in other applied areas (usually under different names).

The idea of writing this book came in the spring of 2002, when I taught basic non-life insurance mathematics at the University of Copenhagen. My handwritten notes were not very much appreciated by the students, and so I decided to come up with some lecture notes for the next course given in spring, 2003. This book is an extended version of those notes and the associated weekly exercises. I have also added quite a few computer graphics to the text. Graphs help one to understand and digest the theory much easier than formulae and proofs. In particular, computer simulations illustrate where the limits of the theory actually are.

When one writes a book, one uses the experience and knowledge of generations of mathematicians without being directly aware of it. Ole Hesselager's 1998 notes and exercises for the basic course on non-life insurance at the Laboratory of Actuarial Mathematics in Copenhagen were a guideline to the content of this book. I also benefitted from the collective experience of writing EKM [29]. The knowledgeable reader will see a few parallels between the two books. However, this book is an *introduction* to non-life insurance, whereas EKM assume that the reader is familiar with the basics of this theory and also explores various other topics of applied probability theory. After having read this book, the reader will be ready for EKM. Another influence has been Sid Resnick's enjoyable book about Happy Harry [65]. I admit that some of the mathematical taste of that book has infected mine; the interested reader will find a wealth of applied stochastic process theory in [65] which goes far beyond the scope of this book.

The choice of topics presented in this book has been dictated, on the one hand, by personal taste and, on the other hand, by some practical considerations. This course is the basis for other courses in the curriculum of the Danish actuarial education and therefore it has to cover a certain variety of topics. This education is in agreement with the Groupe Consultatif requirements, which are valid in most European countries.

As regards personal taste, I very much focused on methods and ideas which, in one way or other, are related to renewal theory and point processes. I am in favor of methods where one can see the underlying probabilistic structure without big machinery or analytical tools. This helps one to strengthen intuition. Analytical tools are like modern cars, whose functioning one cannot understand; one only finds out when they break down. Martingale and Markov process theory do not play an important role in this text. They are acting somewhere in the background and are not especially emphasized, since it is the author's opinion that they are not really needed for an introduction to non-life insurance mathematics. Clearly, one has to pay a price for this approach: lack of elegance in some proofs, but with elegance it is very much like with modern cars.

According to the maxim that non-Bayesians have more fun, Bayesian ideas do not play a major role in this text. Part II on experience rating is therefore rather short, but self-contained. Its inclusion is caused by the practical reasons mentioned above but it also pays respect to the influential contributions of Hans Bühlmann to modern insurance mathematics.

Some readers might miss a chapter on the interplay of insurance and finance, which has been an open subject of discussion for many years. There is no doubt that the modern actuary should be educated in modern financial mathematics, but that requires stochastic calculus and continuous-time martingale theory, which is far beyond the scope of this book. There exists a vast specialized literature on financial mathematics. This theory has dictated most of the research on financial products in insurance. To the best of the author's knowledge, there is no part of insurance mathematics which deals with the pricing and hedging of insurance products by techniques and approaches genuinely different from those of financial mathematics.

It is a pleasure to thank my colleagues and students at the Laboratory of Actuarial Mathematics in Copenhagen for their support. Special thanks go to Jeffrey Collamore, who read much of this text and suggested numerous improvements upon my German way of writing English. I am indebted to Catriona Byrne from Springer-Verlag for professional editorial help.

If this book helps to change the perception that non-life insurance mathematics has nothing to offer but boring calculations, its author has achieved his objective.

Thomas Mikosch   Copenhagen, September 2003

Guidelines to the Reader

This book grew out of an introductory course on non-life insurance, which I taught several times at the Laboratory of Actuarial Mathematics of the University of Copenhagen. This course was given at the third year of the actuarial studies which, together with an introductory course on life insurance, courses on law and accounting, and bachelor projects on life and non-life insurance, leads to the Bachelor's degree in Actuarial Mathematics. This programme has been successfully composed and applied in the 1990s by Ragnar Norberg and his colleagues. In particular, I have benefitted from the notes and exercises of Ole Hesselager which, in a sense, formed the first step to the construction of this book.

When giving a course for the first time, one is usually faced with the situation that one looks for appropriate teaching material: one browses through the available literature (which is vast in the case of non-life insurance), and soon one realizes that the available texts do not exactly suit one's needs for the course.

> ### What are the prerequisites for this book?

Since the students of the Laboratory of Actuarial Mathematics in Copenhagen have quite a good background in measure theory, probability theory and stochastic processes, it is natural to build a course on non-life insurance based on knowledge of these theories. In particular, the theory of stochastic processes and applied probability theory (which insurance mathematics is a part of) have made significant progress over the last 50 years, and therefore it seems appropriate to use these tools even in an introductory course.

On the other hand, the level of this course is not too advanced. For example, martingale and Markov process theory are avoided as much as possible and so are many analytical tools such as Laplace-Stieltjes transforms; these notions only appear in the exercises or footnotes. Instead I focused on a more intuitive probabilistic understanding of the risk and total claim amount processes and their underlying random walk structure. A random walk is one of the simplest stochastic processes and allows in many cases for explicit calculations of distributions and their characteristics. If one goes this way, one essentially walks along the path of renewal and point process theory. However, renewal theory will not be stressed too much, and only some of the essential tools such as the key renewal theorem will be explained at an informal level. Point process theory will be used indirectly at many places, in particular, in the section on the Poisson process, but also in this case the discussion will not go too far; the notion of a random measure will be mentioned but not really needed for the understanding of the succeeding sections and chapters.

Summarizing the above, the reader of this book should have a good background in probability and measure theory and in stochastic processes. Measure theoretic arguments can sometimes be replaced by intuitive arguments, but measure theory will make it easier to get through the chapters of this book.

For whom is this book written?

The book is primarily written for the undergraduate student who wants to learn about some fundamental results in non-life insurance mathematics by using the theory of stochastic processes. One of the differences from other texts of this kind is that I have tried to express most of the theory in the language of stochastic processes. As a matter of fact, Filip Lundberg and Harald Cramér — two pioneers in actuarial mathematics — have worked in exactly this spirit: the insurance business in its parts is described as a continuous-time stochastic process. This gives a more complex view of insurance mathematics and allows one to apply recent results from the theory of stochastic processes.

A widespread opinion about insurance mathematics (at least among mathematicians) is that it is a rather dry and boring topic since one only calculates moments and does not really have any interesting structures. One of the aims of this book is to show that one should not take this opinion at face value and that it is enjoyable to work with the structures of non-life insurance mathematics. Therefore the present text can be interesting also for those who do not necessarily wish to spend the rest of their lives in an insurance company. The reader of this book could be a student in any field of applied mathematics or statistics, a physicist or an engineer who wants to learn about applied stochastic models such as the Poisson, compound Poisson and renewal processes. These processes lie at the heart of this book and are fundamental in many other areas of applied probability theory, such as renewal theory, queuing, stochastic networks, and point process theory. The chapters of this book touch on more general topics than insurance mathematics. The interested reader will find discussions about more advanced topics, with a list of relevant references, showing that insurance mathematics is not a closed world but open to other fields of applied probability theory, stochastic processes and statistics.

How should you read this book?

Part I deals with collective risk models, i.e., models which describe the evolution of an insurance portfolio as a mechanism, where claims and premiums have to be balanced in order to avoid ruin. Part II studies the individual policies and gives advice about how much premium should be charged depending on the policy experience represented by the claim data. There is little theoretical overlap of these two parts; the models and the mathematical tools are completely different.

The core material (and the more interesting one from the author's point of view, since it uses genuine stochastic process theory) is contained in Part I. It is built up in an hierarchical way. You cannot start with Chapter 4 on ruin theory without having understood Chapter 2 on claim number processes.

Chapter 1 introduces the basic model of collective risk theory, combining claim sizes and claim arrival times. The claim number process, i.e., the counting process of the claim arrival times, is one of the main objects of interest in this book. It is dealt with in Chapter 2, where three major claim number processes are introduced: the Poisson process (Section 2.1), the renewal process (Section 2.2) and the mixed Poisson process (Section 2.3). Most of the material of these sections is relevant for the understanding of the remaining sections. However, some of the sections contain informal discussions (for example, about the generalized Poisson process or renewal theory), which can be skipped on first reading; only a few facts of those sections will be used later. The discussions at an informal level are meant as appetizers to make the reader curious and to invite him/her to learn about more advanced probabilistic structures.

Chapter 3 studies the total claim amount process, i.e., the process of the aggregated claim sizes in the portfolio as a function of time. The order of magnitude of this object is of main interest, since it tells one how much premium should be charged in order to avoid ruin. Section 3.1 gives some quantitative measures for the order of magnitude of the total claim amount. Realistic claim size distributions are discussed in Section 3.2. In particular, we stress the notion of heavy-tailed distribution, which lies at the heart of (re)insurance and addresses how large claims or the largest claim can be modeled in an appropriate way. Over the last 30 years we have experienced major man-made and natural catastrophes; see Table 3.2.18, where the largest insurance losses are reported. They challenge the insurance industry, but they also call for improved mathematical modeling. In Section 3.2 we further discuss some exploratory statistical tools and illustrate them with real-life and simulated insurance data. Much of the material of this section is informal and the interested reader is again referred to more advanced literature which might give answers to the questions which arose in the process of reading. In Section 3.3 we touch upon the problem of how one can calculate or approximate the distribution of the total claim amount. Since this is a difficult and complex matter we cannot come up with complete solutions. We rather focus on one of the numerical methods for calculating this distribution, and then we give informal discussions of methods which are based on approximations or simulations. These are quite specific topics and therefore their space is limited in this book. The final Section 3.4 on reinsurance treaties introduces basic notions of the reinsurance language and discusses their relation to the previously developed theory.

Chapter 4 deals with one of the highlights of non-life insurance mathematics: the probability of ruin of a portfolio. Since the early work by Lundberg [55] and Cramér [23], this part has been considered a jewel of the theory. It is rather demanding from a mathematical point of view. On the other hand, the reader learns how various useful concepts of applied probability theory (such as renewal theory, Laplace-Stieltjes transforms, integral equations) enter to solve this complicated problem. Section 4.1 gives a gentle introduction to the topic "ruin". The famous results of Lundberg and Cramér on the order of magnitude of the ruin probability are formulated and proved in Section 4.2. The Cramér result, in particular, is perhaps the most challenging mathematical result of this book. We prove it in detail; only at a few spots do we need to borrow some more advanced tools from renewal theory. Cramér's theorem deals with ruin for the small claim case. We also prove the corresponding result for the large claim case, where one very large claim can cause ruin spontaneously.

As mentioned above, Part II deals with models for the individual policies. Chapters 5 and 6 give a brief introduction to experience rating: how much premium should be charged for a policy based on the claim history? In these two chapters we introduce three major models (heterogeneity, Bühlmann, Bühlmann-Straub) in order to describe the dependence of the claim structure inside a policy and across the policies. Based on these models, we discuss classical methods in order to determine a premium for a policy by taking into account the claim history and the overall portfolio experience (credibility theory). Experience rating and credibility theory are classical and influential parts of non-life insurance mathematics. They do not require genuine techniques from stochastic process theory, but they are nevertheless quite demanding: the proofs are quite technical.

It is recommended that the reader who wishes to be successful should solve the exercises, which are collected at the end of each section; they are an integral part of this course. Moreover, some of the proofs in the sections are only sketched and the reader is recommended to complete them. The exercises also give some guidance to the solution of these problems.

At the end of this book you will know about the fundamental models of non-life insurance mathematics and about applied stochastic processes. Then you may want to know more about stochastic processes in general and insurance models in particular. At the end of the sections and sometimes at suitable spots in the text you will find references to more advanced literature. They can be useful for the continuation of your studies.

> **You are now ready to start. Good luck!**

Part I

Collective Risk Models

1

The Basic Model

In 1903 the Swedish actuary Filip Lundberg [55] laid the foundations of modern risk theory. *Risk theory* is a synonym for non-life insurance mathematics, which deals with the modeling of claims that arrive in an insurance business and which gives advice on how much premium has to be charged in order to avoid bankruptcy (ruin) of the insurance company.

One of Lundberg's main contributions is the introduction of a simple model which is capable of describing the basic dynamics of a homogeneous insurance portfolio. By this we mean a portfolio of *contracts* or *policies* for similar risks such as car insurance for a particular kind of car, insurance against theft in households or insurance against water damage of one-family homes.

There are three assumptions in the model:

- Claims happen at the times T_i satisfying $0 \leq T_1 \leq T_2 \leq \cdots$. We call them *claim arrivals* or *claim times* or *claim arrival times* or, simply, *arrivals*.
- The ith claim arriving at time T_i causes the *claim size* or *claim severity* X_i. The sequence (X_i) constitutes an iid sequence of non-negative random variables.
- The claim size process (X_i) and the claim arrival process (T_i) are *mutually independent*.

The iid property of the claim sizes, X_i, reflects the fact that there is a *homogeneous probabilistic structure* in the portfolio. The assumption that claim sizes and claim times be independent is very natural from an intuitive point of view. But the independence of claim sizes and claim arrivals also makes the life of the mathematician much easier, i.e., this assumption is made for mathematical convenience and tractability of the model.

Now we can define the *claim number process*

$$N(t) = \#\{i \geq 1 : T_i \leq t\}, \quad t \geq 0,$$

i.e., $N = (N(t))_{t \geq 0}$ is a counting process on $[0, \infty)$: $N(t)$ is the number of the claims which occurred by time t.

The object of main interest from the point of view of an insurance company is the *total claim amount process* or *aggregate claim amount process*:[1]

$$S(t) = \sum_{i=1}^{N(t)} X_i = \sum_{i=1}^{\infty} X_i I_{[0,t]}(T_i), \quad t \geq 0.$$

The process $S = (S(t))_{t \geq 0}$ is a *random partial sum process* which refers to the fact that the deterministic index n of the partial sums $S_n = X_1 + \cdots + X_n$ is replaced by the random variables $N(t)$:

$$S(t) = X_1 + \cdots + X_{N(t)} = S_{N(t)}, \quad t \geq 0.$$

It is also often called a *compound (sum) process*. We will observe that the total claim amount process S shares various properties with the partial sum process. For example, asymptotic properties such as the central limit theorem and the strong law of large numbers are analogous for the two processes; see Section 3.1.2.

In Figure 1.0.1 we see a sample path of the process N and the corresponding sample path of the compound sum process S. Both paths jump at the same times T_i: by 1 for N and by X_i for S.

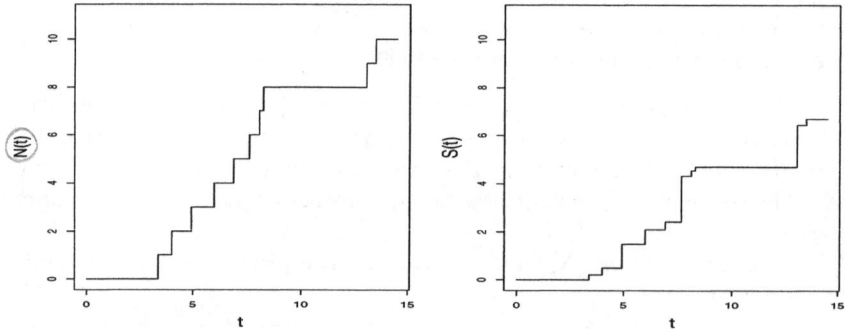

Figure 1.0.1 *A sample path of the claim arrival process N (left) and of the corresponding total claim amount process S (right). Mind the difference of the jump sizes!*

One would like to solve the following problems by means of insurance mathematical methods:

---

[1] Here and in what follows, $\sum_{i=1}^{0} a_i = 0$ for any real a_i and I_A is the indicator function of any set A: $I_A(x) = 1$ if $x \in A$ and $I_A(x) = 0$ if $x \notin A$.

- Find sufficiently realistic, but simple,[2] probabilistic models for S and N. This means that we have to specify the distribution of the claim sizes X_i and to introduce models for the claim arrival times T_i. The discrepancy between "realistic" and "simple" models is closely related to the question to which extent a mathematical model can describe the complicated dynamics of an insurance portfolio without being mathematically intractable.

- Determine the theoretical properties of the stochastic processes S and N. Among other things, we are interested in the distributions of S and N, their distributional characteristics such as the moments, the variance and the dependence structure. We will study the asymptotic behavior of $N(t)$ and $S(t)$ for large t and the average behavior of N and S in the interval $[0,t]$. To be more specific, we will give conditions under which the strong law of large numbers and the central limit theorem hold for S and N.

- Give simulation procedures for the processes N and S. Simulation methods have become more and more popular over the last few years. In many cases they have replaced rigorous probabilistic and/or statistical methods. The increasing power of modern computers allows one to simulate various scenarios of possible situations an insurance business might have to face in the future. This does not mean that no theory is needed any more. On the contrary, simulation generally must be based on probabilistic models for N and S; the simulation procedure itself must exploit the theoretical properties of the processes to be simulated.

- Based on the theoretical properties of N and S, give advice how to choose a premium in order to cover the claims in the portfolio, how to build reserves, how to price insurance products, etc.

Although statistical inference on the processes S and N is utterly important for the insurance business, we do not address this aspect in a rigorous way. The statistical analysis of insurance data is not different from standard statistical methods which have been developed for iid data and for counting processes. Whereas there exist numerous monographs dealing with the inference of iid data, books on the inference of counting processes are perhaps less known. We refer to the book by Andersen et al. [2] for a comprehensive treatment.

We start with the extensive Chapter 2 on the modeling of the claim number process N. The process of main interest is the *Poisson process*. It is treated in Section 2.1. The Poisson process has various attractive theoretical properties which have been collected for several decades. Therefore it is not surprising that it made its way into insurance mathematics from the very beginning, starting with Lundberg's thesis [55]. Although the Poisson process is perhaps not the most realistic process when it comes to fitting real-life claim arrival times, it is kind of a benchmark process. Other models for N are modifications of the Poisson process which yield greater flexibility in one way or the other.

---

[2] This requirement is in agreement with Einstein's maxim "as simple as possible, but not simpler".

This concerns the *renewal process* which is considered in Section 2.2. It allows for more flexibility in choosing the distribution of the inter-arrival times $T_i - T_{i-1}$. But one has to pay a price: in contrast to the Poisson process when $N(t)$ has a Poisson distribution for every t, this property is in general not valid for a renewal process. Moreover, the distribution of $N(t)$ is in general not known. Nevertheless, the study of the renewal process has led to a strong mathematical theory, the so-called *renewal theory*, which allows one to make quite precise statements about the expected claim number $EN(t)$ for large t. We sketch renewal theory in Section 2.2.2 and explain what its purpose is without giving all mathematical details, which would be beyond the scope of this text. We will see in Section 4.2.2 on ruin probabilities that the so-called *renewal equation* is a very powerful tool which gives us a hand on measuring the *probability of ruin* in an insurance portfolio. A third model for the claim number process N is considered in Section 2.3: the *mixed Poisson process*. It is another modification of the Poisson process. By randomization of the parameters of a Poisson process ("mixing") one obtains a class of processes which exhibit a much larger variety of sample paths than for the Poisson or the renewal processes. We will see that the mixed Poisson process has some distributional properties which completely differ from the Poisson process.

After the extensive study of the claim number process we focus in Chapter 3 on the theoretical properties of the *total claim amount process S*. We start in Section 3.1 with a description of the order of magnitude of $S(t)$. Results include the mean and the variance of $S(t)$ (Section 3.1.1) and asymptotic properties such as the strong law of large numbers and the central limit theorem for $S(t)$ as $t \to \infty$ (Section 3.1.2). We also discuss classical premium calculation principles (Section 3.1.3) which are rules of thumb for how large the premium in a portfolio should be in order to avoid ruin. These principles are consequences of the theoretical results on the growth of $S(t)$ for large t. In Section 3.2 we hint at realistic claim size distributions. In particular, we focus on *heavy-tailed* claim size distributions and study some of their theoretical properties. Distributions with regularly varying tails and subexponential distributions are introduced as *the* natural classes of distributions which are capable of describing large claim sizes. Section 3.3 continues with a study of the distributional characteristics of $S(t)$. We show some nice closure properties which certain total claim amount models ("mixture distributions") obey; see Section 3.3.1. We also show the surprising result that a disjoint decomposition of time and/or claim size space yields independent total claim amounts on the different pieces of the partition; see Section 3.3.2. Then various exact (numerical; see Section 3.3.3) and approximate (Monte Carlo, bootstrap, central limit theorem based; see Section 3.3.4) methods for determining the distribution of $S(t)$, their advantages and drawbacks are discussed. Finally, in Section 3.4 we give an introduction to reinsurance treaties and show the link to previous theory.

A major building block of classical risk theory is devoted to the *probability of ruin*; see Chapter 4. It is a global measure of the risk one encounters in a portfolio over a long time horizon. We deal with the classical small claim case and give the celebrated estimates of Cramér and Lundberg (Sections 4.2.1 and 4.2.2). These results basically say that ruin is very unlikely for small claim sizes. In contrast to the latter results, the large claim case yields completely different results: ruin is not unlikely; see Section 4.2.4.

2

Models for the Claim Number Process

2.1 The Poisson Process

In this section we consider the most common claim number process: the *Poisson process*. It has very desirable theoretical properties. For example, one can derive its finite-dimensional distributions explicitly. The Poisson process has a long tradition in applied probability and stochastic process theory. In his 1903 thesis, Filip Lundberg already exploited it as a model for the claim number process N. Later on in the 1930s, Harald Cramér, the famous Swedish statistician and probabilist, extensively developed collective risk theory by using the total claim amount process S with arrivals T_i which are generated by a Poisson process. For historical reasons, but also since it has very attractive mathematical properties, the Poisson process plays a central role in insurance mathematics.

Below we will give a definition of the Poisson process, and for this purpose we now introduce some notation. For any real-valued function f on $[0, \infty)$ we write

$$f(s, t] = f(t) - f(s), \quad 0 \le s < t < \infty.$$

Recall that an integer-valued random variable M is said to have a Poisson distribution with parameter $\lambda > 0$ ($M \sim \text{Pois}(\lambda)$) if it has distribution

$$P(M = k) = e^{-\lambda} \frac{\lambda^k}{k!}, \quad k = 0, 1, \ldots.$$

We say that the random variable $M = 0$ a.s. has a Pois(0) distribution. Now we are ready to define the *Poisson process*.

Definition 2.1.1 (Poisson process)
A stochastic process $N = (N(t))_{t \ge 0}$ is said to be a Poisson process if the following conditions hold:

(1) The process starts at zero: $N(0) = 0$ a.s.

(2) *The process has independent increments: for any t_i, $i = 0, \ldots, n$, and $n \geq 1$ such that $0 = t_0 < t_1 < \cdots < t_n$, the increments $N(t_{i-1}, t_i]$, $i = 1, \ldots, n$, are mutually independent.*

(3) *There exists a non-decreasing right-continuous function $\mu : [0, \infty) \to [0, \infty)$ with $\mu(0) = 0$ such that the increments $N(s, t]$ have a Poisson distribution $\mathrm{Pois}(\mu(s, t])$. We call μ the mean value function of N.*

(4) *With probability 1, the sample paths $(N(t, \omega))_{t \geq 0}$ of the process N are right-continuous for $t \geq 0$ and have limits from the left for $t > 0$. We say that N has càdlàg (continue à droite, limites à gauche) sample paths.*

We continue with some comments on this definition and some immediate consequences.

We know that a Poisson random variable M has the rare property that

$$\lambda = EM = \mathrm{var}(M),$$

i.e., it is determined only by its mean value (= variance) if the distribution is specified as Poisson. The definition of the Poisson process essentially says that, in order to determine the distribution of the Poisson process N, it suffices to know its mean value function. The mean value function μ can be considered as an inner clock or *operational time* of the counting process N. Depending on the magnitude of $\mu(s, t]$ in the interval $(s, t]$, $s < t$, it determines how large the random increment $N(s, t]$ is.

Since $N(0) = 0$ a.s. and $\mu(0) = 0$,

$$N(t) = N(t) - N(0) = N(0, t] \sim \mathrm{Pois}(\mu(0, t]) = \mathrm{Pois}(\mu(t)).$$

We know that the distribution of a stochastic process (in the sense of Kolmogorov's consistency theorem[1]) is determined by its finite-dimensional distributions. The finite-dimensional distributions of a Poisson process have a rather simple structure: for $0 = t_0 < t_1 < \cdots < t_n < \infty$,

$$(N(t_1), N(t_2), \ldots, N(t_n)) =$$

$$\left(N(t_1), N(t_1) + N(t_1, t_2], N(t_1) + N(t_1, t_2] + N(t_2, t_3], \ldots, \sum_{i=1}^{n} N(t_{i-1}, t_i] \right).$$

where any of the random variables on the right-hand side is Poisson distributed. The independent increment property makes it easy to work with the finite-dimensional distributions of N: for any integers $k_i \geq 0$, $i = 1, \ldots, n$,

---

[1] Two stochastic processes on the real line have the same distribution in the sense of Kolmogorov's consistency theorem (cf. Rogers and Williams [66], p. 123, or Billingsley [13], p. 510) if their finite-dimensional distributions coincide. Here one considers the processes as random elements with values in the product space $\mathbb{R}^{[0,\infty)}$ of real-valued functions on $[0, \infty)$, equipped with the σ-field generated by the cylinder sets of $\mathbb{R}^{[0,\infty)}$.

$$P(N(t_1) = k_1, N(t_2) = k_1 + k_2, \ldots, N(t_n) = k_1 + \cdots + k_n)$$

$$= P(N(t_1) = k_1, N(t_1, t_2] = k_2, \ldots, N(t_{n-1}, t_n] = k_n)$$

$$= e^{-\mu(t_1)} \frac{(\mu(t_1))^{k_1}}{k_1!} e^{-\mu(t_1,t_2]} \frac{(\mu(t_1,t_2])^{k_2}}{k_2!} \cdots e^{-\mu(t_{n-1},t_n]} \frac{(\mu(t_{n-1},t_n])^{k_n}}{k_n!}$$

$$= e^{-\mu(t_n)} \frac{(\mu(t_1))^{k_1}}{k_1!} \frac{(\mu(t_1,t_2])^{k_2}}{k_2!} \cdots \frac{(\mu(t_{n-1},t_n])^{k_n}}{k_n!} \, .$$

The càdlàg property is nothing but a standardization property and of purely mathematical interest which, among other things, ensures the measurability property of the stochastic process N in certain function spaces.[2] As a matter of fact, it is possible to show that one can define a process N on $[0, \infty)$ satisfying properties (1)-(3) of the Poisson process and having sample paths which are left-continuous and have limits from the right.[3] Later, in Section 2.1.4, we will give a constructive definition of the Poisson process. That version will automatically be càdlàg.

2.1.1 The Homogeneous Poisson Process, the Intensity Function, the Cramér-Lundberg Model

The most popular Poisson process corresponds to the case of a linear mean value function μ:

$$\mu(t) = \lambda t, \quad t \geq 0,$$

for some $\lambda > 0$. A process with such a mean value function is said to be *homogeneous*, *inhomogeneous* otherwise. The quantity λ is the *intensity* or *rate* of the homogeneous Poisson process. If $\lambda = 1$, N is called *standard homogeneous Poisson process*.

More generally, we say that N has an *intensity function* or *rate function* λ if μ is absolutely continuous, i.e., for any $s < t$ the increment $\mu(s, t]$ has representation

$$\mu(s, t] = \int_s^t \lambda(y) \, dy, \quad s < t,$$

for some non-negative measurable function λ. A particular consequence is that μ is a continuous function.

We mentioned that μ can be interpreted as operational time or inner clock of the Poisson process. If N is homogeneous, time evolves linearly: $\mu(s, t] = \mu(s + h, t + h]$ for any $h > 0$ and $0 \leq s < t < \infty$. Intuitively, this means that

---

[2] A suitable space is the Skorokhod space $\mathbb{D}$ of càdlàg functions on $[0, \infty)$; cf. Billingsley [12].

[3] See Chapter 2 in Sato [71].

claims arrive roughly uniformly over time. We will see later, in Section 2.1.6, that this intuition is supported by the so-called *order statistics property* of a Poisson process. If N has non-constant intensity function λ time "slows down" or "speeds up" according to the magnitude of $\lambda(t)$. In Figure 2.1.2 we illustrate this effect for different choices of λ. In an insurance context, non-constant λ may refer to seasonal effects or trends. For example, in Denmark more car accidents happen in winter than in summer due to bad weather conditions. Trends can, for example, refer to an increasing frequency of (in particular, large) claims over the last few years. Such an effect has been observed in windstorm insurance in Europe and is sometimes mentioned in the context of climate change. Table 3.2.18 contains the largest insurance losses occurring in the period 1970-2002: it is obvious that the arrivals of the largest claim sizes cluster towards the end of this time period. We also refer to Section 2.1.7 for an illustration of seasonal and trend effects in a real-life claim arrival sequence.

A homogeneous Poisson process with intensity λ has

(1) càdlàg sample paths,
(2) starts at zero,
(3) has independent and *stationary* increments,
(4) $N(t)$ is Pois(λt) distributed for every $t > 0$.

Stationarity of the increments refers to the fact that for any $0 \le s < t$ and $h > 0$,

$$N(s,t] \stackrel{d}{=} N(s+h, t+h] \sim \text{Pois}(\lambda\,(t-s))\,,$$

i.e., the Poisson parameter of an increment only depends on the length of the interval, not on its location.

A process on $[0, \infty)$ with properties (1)-(3) is called a *Lévy process*. The homogeneous Poisson process is one of the prime examples of Lévy processes with applications in various areas such as queuing theory, finance, insurance, stochastic networks, to name a few. Another prime example of a Lévy process is *Brownian motion B*. In contrast to the Poisson process, which is a pure jump process, Brownian motion has continuous sample paths with probability 1 and its increments $B(s,t]$ are normally N($0, \sigma^2\,(t-s)$) distributed for some $\sigma > 0$. Brownian motion has a multitude of applications in physics and finance, but also in insurance mathematics. Over the last 30 years, Brownian motion has been used to model prices of speculative assets (share prices, foreign exchange rates, composite stock indices, etc.).

Finance and insurance have been merging for many years. Among other things, insurance companies invest in financial derivatives (options, futures, etc.) which are commonly modeled by functions of Brownian motion such as solutions to stochastic differential equations. If one wants to take into account jump characteristics of real-life financial/insurance phenomena, the Poisson

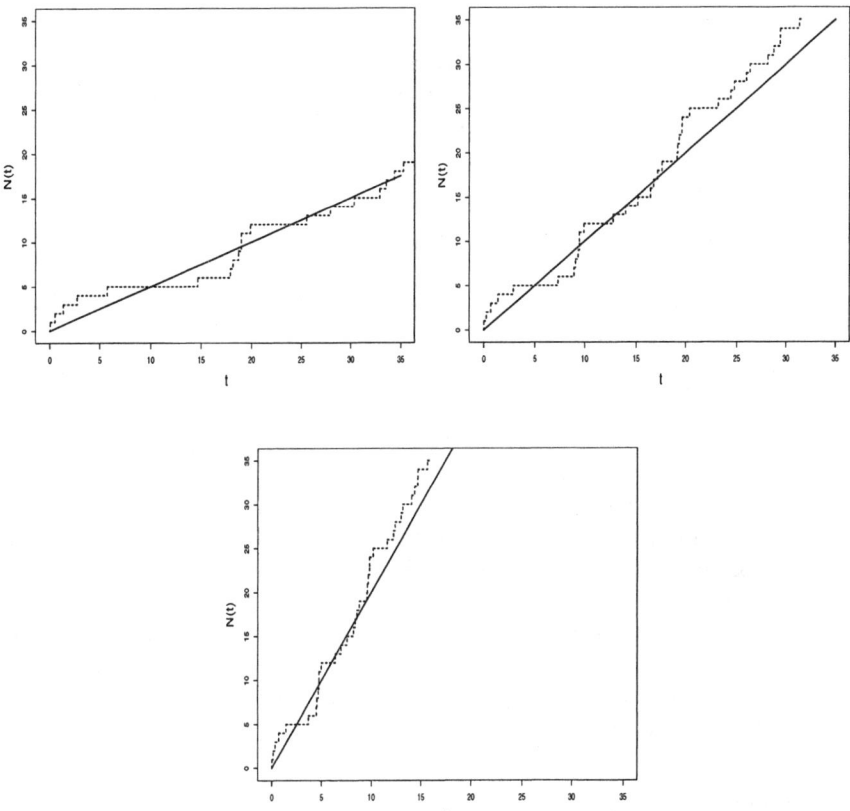

Figure 2.1.2 *One sample path of a Poisson process with intensity* 0.5 *(top left)*, 1 *(top right) and* 2 *(bottom). The straight lines indicate the corresponding mean value functions. For* $\lambda = 0.5$ *jumps occur less often than for the standard homogeneous Poisson process, whereas they occur more often when* $\lambda = 2$.

process, or one of its many modifications, in combination with Brownian motion, offers the opportunity to model financial/insurance data more realistically. In this course, we follow the classical tradition of non-life insurance, where Brownian motion plays a less prominent role. This is in contrast to modern life insurance which deals with the inter-relationship of financial and insurance products. For example, unit-linked life insurance can be regarded as classical life insurance which is linked to a financial underlying such as a composite stock index (DAX, S&P 500, Nikkei, CAC40, etc.). Depending on the performance of the underlying, the policyholder can gain an additional bonus in excess of the cash amount which is guaranteed by the classical life insurance contracts.

Now we introduce one of the models which will be most relevant through-out this text.

Example 2.1.3 (The Cramér-Lundberg model)
The homogeneous Poisson process plays a major role in insurance mathematics. If we specify the claim number process as a homogeneous Poisson process, the resulting model which combines claim sizes and claim arrivals is called *Cramér-Lundberg model* :

- Claims happen at the arrival times $0 \leq T_1 \leq T_2 \leq \cdots$ of a homogeneous Poisson process $N(t) = \#\{i \geq 1 : T_i \leq t\}$, $t \geq 0$.
- The ith claim arriving at time T_i causes the claim size X_i. The sequence (X_i) constitutes an iid sequence of non-negative random variables.
- The sequences (T_i) and (X_i) are independent. In particular, N and (X_i) are independent.

The total claim amount process S in the Cramér-Lundberg model is also called a *compound Poisson process*.

The Cramér-Lundberg model is one of the most popular and useful models in non-life insurance mathematics. Despite its simplicity it describes some of the essential features of the total claim amount process which is observed in reality.

We mention in passing that the total claim amount process S in the Cramér-Lundberg setting is a process with independent and stationary increments, starts at zero and has càdlàg sample paths. It is another important example of a Lévy process. Try to show these properties!   □

Comments

The reader who wants to learn about Lévy processes is referred to Sato's monograph [71]. For applications of Lévy processes in different areas, see the recent collection of papers edited by Barndorff-Nielsen et al. [9]. Rogers and Williams [66] can be recommended as an introduction to Brownian motion, its properties and related topics such as stochastic differential equations. For an elementary introduction, see Mikosch [57].

2.1.2 The Markov Property

Poisson processes constitute one particular class of *Markov processes* on $[0, \infty)$ with state space $\mathbb{N}_0 = \{0, 1, \ldots\}$. This is a simple consequence of the independent increment property. It is left as an exercise to verify the *Markov property*, i.e., for any $0 = t_0 < t_1 < \cdots < t_n$ and non-decreasing natural numbers $k_i \geq 0$, $i = 1, \ldots, n$,

$$P(N(t_n) = k_n \mid N(t_1) = k_1, \ldots, N(t_{n-1}) = k_{n-1})$$

$$= P(N(t_n) = k_n \mid N(t_{n-1}) = k_{n-1}).$$

Markov process theory does not play a prominent role in this course,[4] in contrast to a course on modern life insurance mathematics, where Markov models are fundamental.[5] However, the *intensity function of a Poisson process* N has a nice interpretation as the *intensity function of the Markov process* N. Before we make this statement precise, recall that the quantities

$$p_{k,k+h}(s,t) = P(N(t) = k + h \mid N(s) = k) = P(N(t) - N(s) = h),$$

$$0 \le s < t, \quad k, h \in \mathbb{N}_0,$$

are called the *transition probabilities* of the Markov process N with state space $\mathbb{N}_0$. Since a.e. path $(N(t, \omega))_{t \ge 0}$ increases with probability 1 (verify this), one only needs to consider transitions of the Markov process N from k to $k + h$ for $h \ge 0$. The transition probabilities are closely related to the *intensities* which are given as the limits

$$\lambda_{k,k+h}(t) = \lim_{s \downarrow 0} \frac{p_{k,k+h}(t, t + s)}{s},$$

provided they and their analogs from the left exist, are finite and coincide. From the theory of stochastic processes, we know that the intensities and the initial distribution of a Markov process determine the distribution of this Markov process.[6]

Proposition 2.1.4 (Relation of the intensity function of the Poisson process and its Markov intensities)
Consider a Poisson process $N = (N(t))_{t \ge 0}$ which has a continuous intensity function λ on $[0, \infty)$. Then, for $k \ge 0$,

$$\lambda_{k,k+h}(t) = \begin{cases} \lambda(t) & \text{if } h = 1, \\ 0 & \text{if } h > 1. \end{cases}$$

In words, the intensity function $\lambda(t)$ of the Poisson process N is nothing but the intensity of the Markov process N for the transition from state k to state $k + 1$. The proof of this result is left as an exercise.

The intensity function of a Markov process is a quantitative measure of the likelihood that the Markov process N jumps in a small time interval. An immediate consequence of Proposition 2.1.4 is that is it is very unlikely that a Poisson process with continuous intensity function λ has jump sizes larger

---

[4] It is, however, no contradiction to say that almost all stochastic models in this course have a Markov structure. But we do not emphasize this property.

[5] See for example Koller [52].

[6] We leave this statement as vague as it is. The interested reader is, for example, referred to Resnick [65] or Rogers and Williams [66] for further reading on Markov processes.

than 1. Indeed, consider the probability that N has a jump greater than 1 in the interval $(t, t+s]$ for some $t \geq 0$, $s > 0$:[7]

$$P(N(t, t+s] \geq 2) = 1 - P(N(t, t+s] = 0) - P(N(t, t+s] = 1)$$

$$= 1 - e^{-\mu(t, t+s]} - \mu(t, t+s] e^{-\mu(t, t+s]}. \qquad (2.1.1)$$

Since λ is continuous,

$$\mu(t, t+s] = \int_t^{t+s} \lambda(y)\, dy = s\,\lambda(t)\,(1 + o(1)) \to 0, \quad \text{as } s \downarrow 0.$$

Moreover, a Taylor expansion yields for $x \to 0$ that $e^x = 1 + x + o(x)$. Thus we may conclude from (2.1.1) that, as $s \downarrow 0$,

$$P(N(t, t+s] \geq 2) = o(\mu(t, t+s]) = o(s). \qquad (2.1.2)$$

It is easily seen that

$$P(N(t, t+s] = 1) = \lambda(t)\, s\, (1 + o(1)). \qquad (2.1.3)$$

Relations (2.1.2) and (2.1.3) ensure that a Poisson process N with continuous intensity function λ is very unlikely to have jump sizes larger than 1. Indeed, we will see in Section 2.1.4 that N has only upward jumps of size 1 with probability 1.

2.1.3 Relations Between the Homogeneous and the Inhomogeneous Poisson Process

The homogeneous and the inhomogeneous Poisson processes are very closely related: we will show in this section that a deterministic time change transforms a homogeneous Poisson process into an inhomogeneous Poisson process, and vice versa.

Let N be a Poisson process on $[0, \infty)$ with mean value function[8] μ. We start with a standard homogeneous Poisson process $\widetilde{N}$ and define

$$\widehat{N}(t) = \widetilde{N}(\mu(t)), \quad t \geq 0.$$

It is not difficult to see that $\widehat{N}$ is again a Poisson process on $[0, \infty)$. (Verify this! Notice that the càdlàg property of μ is used to ensure the càdlàg property of the sample paths $\widehat{N}(t, \omega)$.) Since

---

[7] Here and in what follows, we frequently use the o-notation. Recall that we write for any real-valued function h, $h(x) = o(1)$ as $x \to x_0 \in [-\infty, \infty]$ if $\lim_{x \to x_0} h(x) = 0$ and we write $h(x) = o(g(x))$ as $x \to x_0$ if $h(x) = g(x)\, o(1)$ for any real-valued function $g(x)$.

[8] Recall that the mean value function of a Poisson process starts at zero, is nondecreasing, right-continuous and finite on $[0, \infty)$. In particular, it is a càdlàg function.

$$\widehat{\mu}(t) = E\widehat{N}(t) = E\widetilde{N}(\mu(t)) = \mu(t)\,, \quad t \geq 0\,,$$

and since the distribution of the Poisson process $\widehat{N}$ is determined by its mean value function $\widehat{\mu}$, it follows that $N \overset{d}{=} \widehat{N}$, where $\overset{d}{=}$ refers to equality of the finite-dimensional distributions of the two processes. Hence the processes $\widehat{N}$ and N are not distinguishable from a probabilistic point of view, in the sense of Kolmogorov's consistency theorem; see the remark on p. 14. Moreover, the sample paths of $\widehat{N}$ are càdlàg as required in the definition of the Poisson process.

Now assume that N has a continuous and increasing mean value function μ. This property is satisfied if N has an a.e. positive intensity function λ. Then the inverse μ^{-1} of μ exists. It is left as an exercise to show that the process $\widetilde{N}(t) = N(\mu^{-1}(t))$ is a standard homogeneous Poisson process on $[0, \infty)$ if $\lim_{t\to\infty} \mu(t) = \infty$.[9]

We summarize our findings.

Proposition 2.1.5 (The Poisson process under change of time)
Let μ be the mean value function of a Poisson process N and $\widetilde{N}$ be a standard homogeneous Poisson process. Then the following statements hold:

(1) *The process $(\widetilde{N}(\mu(t)))_{t\geq 0}$ is Poisson with mean value function μ.*
(2) *If μ is continuous, increasing and $\lim_{t\to\infty} \mu(t) = \infty$ then $(N(\mu^{-1}(t)))_{t\geq 0}$ is a standard homogeneous Poisson process.*

This result, which immediately follows from the definition of a Poisson process, allows one in most cases of practical interest to switch from an inhomogeneous Poisson process to a homogeneous one by a simple time change. In particular, it suggests a straightforward way of simulating sample paths of an inhomogeneous Poisson process N from the paths of a homogeneous Poisson process. In an insurance context, one will usually be faced with inhomogeneous claim arrival processes. The above theory allows one to make an "operational time change" to a homogeneous model for which the theory is more accessible. See also Section 2.1.7 for a real-life example.

2.1.4 The Homogeneous Poisson Process as a Renewal Process

In this section we study the sequence of the arrival times $0 \leq T_1 \leq T_2 \leq \cdots$ of a homogeneous Poisson process with intensity $\lambda > 0$. It is our aim to find a constructive way for determining the sequence of arrivals, which in turn can be used as an alternative definition of the homogeneous Poisson process. This characterization is useful for studying the path properties of the Poisson process or for simulating sample paths.

---

[9] If $\lim_{t\to\infty} \mu(t) = y_0 < \infty$ for some $y_0 > 0$, μ^{-1} is defined on $[0, y_0)$ and $\widetilde{N}(t) = N(\mu^{-1}(t))$ satisfies the properties of a standard homogeneous Poisson process restricted to the interval $[0, y_0)$. In Section 2.1.8 it is explained that such a process can be interpreted as a Poisson process on $[0, y_0)$.

We will show that any homogeneous Poisson process with intensity $\lambda > 0$ has representation

$$N(t) = \#\{i \geq 1 : T_i \leq t\}, \quad t \geq 0, \tag{2.1.4}$$

where

$$T_n = W_1 + \cdots + W_n, \quad n \geq 1, \tag{2.1.5}$$

and (W_i) is an iid exponential $\mathrm{Exp}(\lambda)$ sequence. In what follows, it will be convenient to write $T_0 = 0$. Since the random walk (T_n) with non-negative step sizes W_n is also referred to as *renewal sequence*, a process N with representation (2.1.4)-(2.1.5) for a general iid sequence (W_i) is called a *renewal (counting) process*. We will consider general renewal processes in Section 2.2.

Theorem 2.1.6 (The homogeneous Poisson process as a renewal process)

(1) *The process N given by (2.1.4) and (2.1.5) with an iid exponential $\mathrm{Exp}(\lambda)$ sequence (W_i) constitutes a homogeneous Poisson process with intensity $\lambda > 0$.*
(2) *Let N be a homogeneous Poisson process with intensity λ and arrival times $0 \leq T_1 < T_2 \leq \cdots$. Then N has representation (2.1.4), and (T_i) has representation (2.1.5) for an iid exponential $\mathrm{Exp}(\lambda)$ sequence (W_i).*

Proof. (1) We start with a renewal sequence (T_n) as in (2.1.5) and set $T_0 = 0$ for convenience. Recall the defining properties of a Poisson process from Definition 2.1.1. The property $N(0) = 0$ a.s. follows since $W_1 > 0$ a.s. By construction, a path $(N(t, \omega))_{t \geq 0}$ assumes the value i in $[T_i, T_{i+1})$ and jumps at T_{i+1} to level $i + 1$. Hence the sample paths are càdlàg; cf. p. 14 for a definition.

Next we verify that $N(t)$ is $\mathrm{Pois}(\lambda t)$ distributed. The crucial relationship is given by

$$\{N(t) = n\} = \{T_n \leq t < T_{n+1}\}, \quad n \geq 0. \tag{2.1.6}$$

Since $T_n = W_1 + \cdots + W_n$ is the sum of n iid $\mathrm{Exp}(\lambda)$ random variables it is a well-known property that T_n has a gamma $\Gamma(n, \lambda)$ distribution[10] for $n \geq 1$:

$$P(T_n \leq x) = 1 - e^{-\lambda x} \sum_{k=0}^{n-1} \frac{(\lambda x)^k}{k!}.$$

Hence

$$P(N(t) = n) = P(T_n \leq t) - P(T_{n+1} \leq t) = e^{-\lambda t} \frac{(\lambda t)^n}{n!}.$$

---

[10] You can easily verify that this is the distribution function of a $\Gamma(n, \lambda)$ distribution by taking the first derivative. The resulting probability density has the well-known gamma form $\lambda (\lambda x)^{n-1} e^{-\lambda x}/(n-1)!$. The $\Gamma(n, \lambda)$ distribution for $n \in \mathbb{N}$ is also known as the *Erlang distribution* with parameter (n, λ).

This proves the Poisson property of $N(t)$.

Now we switch to the independent stationary increment property. We use a direct "brute force" method to prove this property. A more elegant way via point process techniques is indicated in Resnick [65], Proposition 4.8.1. Since the case of arbitrarily many increments becomes more involved, we focus on the case of two increments in order to illustrate the method. The general case is analogous but requires some bookkeeping. We focus on the adjacent increments $N(t) = N(0, t]$ and $N(t, t + h]$ for $t, h > 0$. We have to show that for any $k, l \in \mathbb{N}_0$,

$$
\begin{aligned}
q_{k,k+l}(t, t + h) &= P(N(t) = k, N(t, t + h] = l) \\
&= P(N(t) = k)\, P(N(t, t + h] = l) \\
&= P(N(t) = k)\, P(N(h) = l) \\
&= e^{-\lambda(t+h)} \frac{(\lambda t)^k (\lambda h)^l}{k!\, l!}.
\end{aligned}
\tag{2.1.7}
$$

We start with the case $l = 0$, $k \geq 1$; the case $l = k = 0$ being trivial. We make use of the relation

$$
\{N(t) = k, N(t, t + h] = l\} = \{N(t) = k, N(t + h) = k + l\}. \tag{2.1.8}
$$

Then, by (2.1.6) and (2.1.8),

$$
\begin{aligned}
q_{k,k+l}(t, t + h) &= P(T_k \leq t < T_{k+1}, T_k \leq t + h < T_{k+1}) \\
&= P(T_k \leq t, t + h < T_k + W_{k+1}).
\end{aligned}
$$

Now we can use the facts that T_k is $\Gamma(k, \lambda)$ distributed with density $\lambda^k x^{k-1} e^{-\lambda x} / (k - 1)!$ and W_{k+1} is $\mathrm{Exp}(\lambda)$ distributed with density $\lambda e^{-\lambda x}$:

$$
\begin{aligned}
q_{k,k+l}(t, t + h) &= \int_0^t e^{-\lambda z} \frac{\lambda(\lambda z)^{k-1}}{(k-1)!} \int_{t+h-z}^\infty \lambda e^{-\lambda x}\, dx\, dz \\
&= \int_0^t e^{-\lambda z} \frac{\lambda(\lambda z)^{k-1}}{(k-1)!} e^{-\lambda(t+h-z)}\, dz \\
&= e^{-\lambda(t+h)} \frac{(\lambda t)^k}{k!}.
\end{aligned}
$$

For $l \geq 1$ we use another conditioning argument and (2.1.6):

$$
\begin{aligned}
&q_{k,k+l}(t, t + h) \\
&= P(T_k \leq t < T_{k+1}, T_{k+l} \leq t + h < T_{k+l+1}) \\
&= E[I_{\{T_k \leq t < T_{k+1} \leq t+h\}} \\
&\quad P(T_{k+l} - T_{k+1} \leq t + h - T_{k+1} < T_{k+l+1} - T_{k+1} \mid T_k, T_{k+1})].
\end{aligned}
$$

Let N' be an independent copy of N, i.e., $N' \stackrel{d}{=} N$. Appealing to (2.1.6) and the independence of T_{k+1} and $(T_{k+l} - T_{k+1}, T_{k+l+1} - T_{k+1})$, we see that

$$q_{k,k+l}(t, t+h)$$

$$= E[I_{\{T_k \le t < T_{k+1} \le t+h\}} P(N'(t+h-T_{k+1}) = l - 1 \mid T_{k+1})]$$

$$= \int_0^t e^{-\lambda z} \frac{\lambda (\lambda z)^{k-1}}{(k-1)!} \int_{t-z}^{t+h-z} \lambda e^{-\lambda x} P(N(t+h-z-x) = l-1) \, dx \, dz$$

$$= \int_0^t e^{-\lambda z} \frac{\lambda (\lambda z)^{k-1}}{(k-1)!} \int_{t-z}^{t+h-z} \lambda e^{-\lambda x} e^{-\lambda (t+h-z-x)} \frac{(\lambda (t+h-z-x))^{l-1}}{(l-1)!}$$
$$dx \, dz$$

$$= e^{-\lambda (t+h)} \int_0^t \frac{\lambda (\lambda z)^{k-1}}{(k-1)!} \, dz \int_0^h \frac{\lambda (\lambda x)^{l-1}}{(l-1)!} \, dx$$

$$= e^{-\lambda (t+h)} \frac{(\lambda t)^k}{k!} \frac{(\lambda h)^l}{l!} \,.$$

This is the desired relationship (2.1.7). Since

$$P(N(t, t+h] = l) = \sum_{k=0}^{\infty} P(N(t) = k, N(t, t+h] = l) \,,$$

it also follows from (2.1.7) that

$$P(N(t) = k, N(t, t+h] = l) = P(N(t) = k) P(N(h) = l) \,.$$

If you have enough patience prove the analog to (2.1.7) for finitely many increments of N.

(2) Consider a homogeneous Poisson process with arrival times $0 \le T_1 \le T_2 \le \cdots$ and intensity $\lambda > 0$. We need to show that there exist iid exponential $\mathrm{Exp}(\lambda)$ random variables W_i such that $T_n = W_1 + \cdots + W_n$, i.e., we need to show that, for any $0 \le x_1 \le x_2 \le \cdots \le x_n$, $n \ge 1$,

$$P(T_1 \le x_1, \ldots, T_n \le x_n)$$

$$= P(W_1 \le x_1, \ldots, W_1 + \cdots + W_n \le x_n)$$

$$= \int_{w_1=0}^{x_1} \lambda e^{-\lambda w_1} \int_{w_2=0}^{x_2-w_1} \lambda e^{-\lambda w_2} \cdots \int_{w_n=0}^{x_n-w_1-\cdots-w_{n-1}} \lambda e^{-\lambda w_n} \, dw_n \cdots dw_1.$$

The verification of this relation is left as an exercise. Hint: It is useful to exploit the relationship

$$\{T_1 \le x_1, \ldots, T_n \le x_n\} = \{N(x_1) \ge 1, \ldots, N(x_n) \ge n\}$$

for $0 \leq x_1 \leq \cdots \leq x_n$, $n \geq 1$.   □

An important consequence of Theorem 2.1.6 is that the inter-arrival times

$$W_i = T_i - T_{i-1}, \quad i \geq 1,$$

of a homogeneous Poisson process with intensity λ are iid $\mathrm{Exp}(\lambda)$. In particular, $T_i < T_{i+1}$ a.s. for $i \geq 1$, i.e., with probability 1 a homogeneous Poisson
process does not have jump sizes larger than 1. Since by the strong law of
large numbers $T_n/n \overset{\text{a.s.}}{\to} EW_1 = \lambda^{-1} > 0$, we may also conclude that T_n grows
roughly like n/λ, and therefore there are no limit points in the sequence (T_n)
at any finite instant of time. This means that the values $N(t)$ of a homogeneous Poisson process are finite on any finite time interval $[0, t]$.

The Poisson process has many amazing properties. One of them is the
following phenomenon which runs in the literature under the name *inspection
paradox*.

Example 2.1.7 (The inspection paradox)
Assume that you study claims which arrive in the portfolio according to a
homogeneous Poisson process N with intensity λ. We have learned that the
inter-arrival times $W_n = T_n - T_{n-1}$, $n \geq 1$, with $T_0 = 0$, constitute an iid
$\mathrm{Exp}(\lambda)$ sequence. Observe the portfolio at a fixed instant of time t. The last
claim arrived at time $T_{N(t)}$ and the next claim will arrive at time $T_{N(t)+1}$.
Three questions arise quite naturally:

(1) What is the distribution of $B(t) = t - T_{N(t)}$, i.e., the length of the period
    $(T_{N(t)}, t]$ since the last claim occurred?
(2) What is the distribution of $F(t) = T_{N(t)+1} - t$, i.e., the length of the period
    $(t, T_{N(t)+1}]$ until the next claim arrives?
(3) What can be said about the joint distribution of $B(t)$ and $F(t)$?

The quantity $B(t)$ is often referred to as *backward recurrence time* or *age*,
whereas $F(t)$ is called *forward recurrence time*, *excess life* or *residual life*.

Intuitively, since t lies somewhere between two claim arrivals and since the
inter-arrival times are iid $\mathrm{Exp}(\lambda)$, we would perhaps expect that $P(B(t) \leq
x_1) < 1 - \mathrm{e}^{-\lambda x_1}$, $x_1 < t$, and $P(F(t) \leq x_2) < 1 - \mathrm{e}^{-\lambda x_2}$, $x_2 > 0$. However,
these conjectures are not confirmed by calculation of the joint distribution
function of $B(t)$ and $F(t)$ for $x_1, x_2 \geq 0$:

$$G_{B(t),F(t)}(x_1, x_2) = P(B(t) \leq x_1, F(t) \leq x_2).$$

Since $B(t) \leq t$ a.s. we consider the cases $x_1 < t$ and $x_1 \geq t$ separately. We
observe for $x_1 < t$ and $x_2 > 0$,

$$\{B(t) \leq x_1\} = \{t - x_1 \leq T_{N(t)} \leq t\} = \{N(t - x_1, t] \geq 1\},$$

$$\{F(t) \leq x_2\} = \{t < T_{N(t)+1} \leq t + x_2\} = \{N(t, t + x_2] \geq 1\}.$$

Hence, by the independent stationary increments of N,

$$G_{B(t),F(t)}(x_1, x_2) = P\left(N(t - x_1, t] \geq 1, N(t, t + x_2] \geq 1\right)$$

$$= P\left(N(t - x_1, t] \geq 1\right) P\left(N(t, t + x_2] \geq 1\right)$$

$$= \left(1 - e^{-\lambda x_1}\right)\left(1 - e^{-\lambda x_2}\right). \tag{2.1.9}$$

An analogous calculation for $x_1 \geq t$, $x_2 \geq 0$ and (2.1.9) yield

$$G_{B(t),F(t)}(x_1, x_2) = \left[(1 - e^{-\lambda x_1}) I_{[0,t)}(x_1) + I_{[t,\infty)}(x_1)\right]\left(1 - e^{-\lambda x_2}\right).$$

Hence $B(t)$ and $F(t)$ are independent, $F(t)$ is $\mathrm{Exp}(\lambda)$ distributed and $B(t)$ has a truncated exponential distribution with a jump at t:

$$P(B(t) \leq x_1) = 1 - e^{-\lambda x_1}, \quad x_1 < t, \quad \text{and} \quad P(B(t) = t) = e^{-\lambda t}.$$

This means in particular that the forward recurrence time $F(t)$ has the same $\mathrm{Exp}(\lambda)$ distribution as the inter-arrival times W_i of the Poisson process N. This property is closely related to the *forgetfulness* property of the exponential distribution:

$$P(W_1 > x + y \mid W_1 > x) = P(W_1 > x), \quad x, y \geq 0,$$

(Verify the correctness of this relation.) and is also reflected in the independent increment property of the Poisson property. It is interesting to observe that

$$\lim_{t \to \infty} P(B(t) \leq x_1) = 1 - e^{-\lambda x_1}, \quad x_1 > 0.$$

Thus, *in an "asymptotic" sense*, both $B(t)$ and $F(t)$ become independent and are exponentially distributed with parameter λ.

We will return to the forward and backward recurrence times of a general renewal process, i.e., when W_i are not necessarily iid exponential random variables, in Example 2.2.14. □

2.1.5 The Distribution of the Inter-Arrival Times

By virtue of Proposition 2.1.5, an inhomogeneous Poisson process N with mean value function μ can be interpreted as a time changed standard homogeneous Poisson process $\widetilde{N}$:

$$(N(t))_{t \geq 0} \overset{d}{=} (\widetilde{N}(\mu(t)))_{t \geq 0}.$$

In particular, let $(\widetilde{T}_i)$ be the arrival sequence of $\widetilde{N}$ and μ be increasing and continuous. Then the inverse μ^{-1} exists and

$$N'(t) = \#\{i \geq 1 : \widetilde{T}_i \leq \mu(t)\} = \#\{i \geq 1 : \mu^{-1}(\widetilde{T}_i) \leq t\}, \quad t \geq 0,$$

is a representation of N in the sense of identity of the finite-dimensional distributions, i.e., $N \overset{d}{=} N'$. Therefore and by virtue of Theorem 2.1.6 the arrival times of an inhomogeneous Poisson process with mean value function μ have representation

$$T_n = \mu^{-1}(\widetilde{T}_n), \quad \widetilde{T}_n = \widetilde{W}_1 + \cdots + \widetilde{W}_n, \quad n \geq 1, \quad \widetilde{W}_i \text{ iid Exp}(1).$$
(2.1.10)

Proposition 2.1.8 (Joint distribution of arrival/inter-arrival times)
Assume N is a Poisson process on $[0, \infty)$ with a continuous a.e. positive intensity function λ. Then the following statements hold.

(1) *The vector of the arrival times $(T_1, \ldots, T_n)$ has density*

$$f_{T_1,\ldots,T_n}(x_1, \ldots, x_n) = e^{-\mu(x_n)} \prod_{i=1}^{n} \lambda(x_i) \, I_{\{0 < x_1 < \cdots < x_n\}}. \quad (2.1.11)$$

(2) *The vector of inter-arrival times $(W_1, \ldots, W_n) = (T_1, T_2 - T_1, \ldots, T_n - T_{n-1})$ has density*

$$f_{W_1,\ldots,W_n}(x_1, \ldots, x_n) = e^{-\mu(x_1 + \cdots + x_n)} \prod_{i=1}^{n} \lambda(x_1 + \cdots + x_i), \quad x_i \geq 0.$$
(2.1.12)

Proof. Since the intensity function λ is a.e. positive and continuous, $\mu(t) = \int_0^t \lambda(s) \, ds$ is increasing and μ^{-1} exists. Moreover, μ is differentiable, and $\mu'(t) = \lambda(t)$. We make use of these two facts in what follows.
(1) We start with a standard homogeneous Poisson process. Then its arrivals $\widetilde{T}_n$ have representation $\widetilde{T}_n = \widetilde{W}_1 + \cdots + \widetilde{W}_n$ for an iid standard exponential sequence $(\widetilde{W}_i)$. The joint density of $(\widetilde{T}_1, \ldots, \widetilde{T}_n)$ is obtained from the joint density of $(\widetilde{W}_1, \ldots, \widetilde{W}_n)$ via the transformation:

$$(y_1, \ldots, y_n) \overset{S}{\to} (y_1, y_1 + y_2, \ldots, y_1 + \cdots + y_n),$$

$$(z_1, \ldots, z_n) \overset{S^{-1}}{\to} (z_1, z_2 - z_1, \ldots, z_n - z_{n-1}).$$

Note that $\det(\partial S(\mathbf{y})/\partial \mathbf{y}) = 1$. Standard techniques for density transformations (cf. Billingsley [13], p. 229) yield for $0 < x_1 < \cdots < x_n$,

$$f_{\widetilde{T}_1,\ldots,\widetilde{T}_n}(x_1, \ldots, x_n) = f_{\widetilde{W}_1,\ldots,\widetilde{W}_n}(x_1, x_2 - x_1, \ldots, x_n - x_{n-1})$$

$$= e^{-x_1} e^{-(x_2 - x_1)} \cdots e^{-(x_n - x_{n-1})} = e^{-x_n}.$$

Since μ^{-1} exists we conclude from (2.1.10) that for $0 < x_1 < \cdots < x_n$,

$$P(T_1 \leq x_1, \ldots, T_n \leq x_n) = P(\mu^{-1}(\widetilde{T}_1) \leq x_1, \ldots, \mu^{-1}(\widetilde{T}_n) \leq x_n)$$

$$= P(\widetilde{T}_1 \leq \mu(x_1), \ldots, \widetilde{T}_n \leq \mu(x_n))$$

$$= \int_0^{\mu(x_1)} \cdots \int_0^{\mu(x_n)} f_{\tilde{T}_1,\ldots,\tilde{T}_n}(y_1,\ldots,y_n)\,dy_1\cdots dy_n$$

$$= \int_0^{\mu(x_1)} \cdots \int_0^{\mu(x_n)} e^{-y_n} I_{\{y_1 < \cdots < y_n\}}\,dy_1\cdots dy_n\,.$$

Taking partial derivatives with respect to the variables $x_1,\ldots,x_n$ and noticing that $\mu'(x_i) = \lambda(x_i)$, we obtain the desired density (2.1.11).

(2) Relation (2.1.12) follows by an application of the above transformations S and S^{-1} from the density of $(T_1,\ldots,T_n)$:

$$f_{W_1,\ldots,W_n}(w_1,\ldots,w_n) = f_{T_1,\ldots,T_n}(w_1, w_1 + w_2,\ldots, w_1 + \cdots + w_n)\,.$$

□

From (2.1.12) we may conclude that the joint density of $W_1,\ldots,W_n$ can be written as the product of the densities of the W_i's if and only if $\lambda(\cdot) \equiv \lambda$ for some positive constant λ. This means that only in the case of a homogeneous Poisson process are the inter-arrival times $W_1,\ldots,W_n$ independent (and identically distributed). This fact is another property which distinguishes the homogeneous Poisson process within the class of all Poisson processes on $[0,\infty)$.

2.1.6 The Order Statistics Property

In this section we study one of the most important properties of the Poisson process which in a sense characterizes the Poisson process. It is the *order statistics property* which it shares only with the mixed Poisson process to be considered in Section 2.3. In order to formulate this property we first give a well-known result on the distribution of the order statistics

$$X_{(1)} \leq \cdots \leq X_{(n)}$$

of an iid sample $X_1,\ldots,X_n$.

Lemma 2.1.9 (Joint density of order statistics)
If the iid X_i's have density f then the density of the vector $(X_{(1)},\ldots,X_{(n)})$ is given by

$$f_{X_{(1)},\ldots,X_{(n)}}(x_1,\ldots,x_n) = n! \prod_{i=1}^n f(x_i)\, I_{\{x_1 < \cdots < x_n\}}\,.$$

Remark 2.1.10 By construction of the order statistics, the support of the vector $(X_{(1)},\ldots,X_{(n)})$ is the set

$$C_n = \{(x_1,\ldots,x_n) : x_1 \leq \cdots \leq x_n\} \subset \mathbb{R}^n\,,$$

and therefore the density $f_{X_{(1)},\ldots,X_{(n)}}$ vanishes outside C_n. Since the existence of a density of X_i implies that all elements of the iid sample $X_1,\ldots,X_n$ are different a.s., the $\leq$'s in the definition of C_n could be replaced by $<$'s. □

Proof. We start by recalling that the iid sample $X_1, \ldots, X_n$ with common density f has *no ties*. This means that the event

$$\widetilde{\Omega} = \{X_{(1)} < \cdots < X_{(n)}\} = \{X_i \neq X_j \text{ for } 1 \leq i < j \leq n\}$$

has probability 1. It is an immediate consequence of the fact that for $i \neq j$,

$$P(X_i = X_j) = E[P(X_i = X_j \mid X_j)] = \int_{\mathbb{R}} P(X_i = y) f(y) \, dy = 0,$$

since $P(X_i = y) = \int_{\{y\}} f(z) \, dz = 0$. Then

$$1 - P(\widetilde{\Omega}) = P\left(\bigcup_{1 \leq i < j \leq n} \{X_i = X_j\} \right) \leq \sum_{1 \leq i < j \leq n} P(X_i = X_j) = 0.$$

Now we turn to the proof of the statement of the lemma. Let Π_n be the set of the permutations π of $\{1, \ldots, n\}$. Fix the values $x_1 < \cdots < x_n$. Then

$$P\left(X_{(1)} \leq x_1, \ldots, X_{(n)} \leq x_n\right) = P\left(\bigcup_{\pi \in \Pi_n} A_\pi \right), \qquad (2.1.13)$$

where

$$A_\pi = \{X_{\pi(i)} = X_{(i)}, i = 1, \ldots, n\} \cap \widetilde{\Omega} \cap \{X_{\pi(1)} \leq x_1, \ldots, X_{\pi(n)} \leq x_n\}.$$

The identity (2.1.13) means that the ordered sample $X_{(1)} < \cdots < X_{(n)}$ could have come from any of the ordered values $X_{\pi(1)} < \cdots < X_{\pi(n)}$, $\pi \in \Pi_n$, where we also make use of the fact that there are no ties in the sample. Since the A_π's are disjoint,

$$P\left(\bigcup_{\pi \in \Pi_n} A_\pi \right) = \sum_{\pi \in \Pi_n} P(A_\pi).$$

Moreover, since the X_i's are iid,

$$P(A_\pi) = P\left((X_{\pi(1)}, \ldots, X_{\pi(n)}) \in C_n \cap (-\infty, x_1] \times \cdots \times (-\infty, x_n]\right)$$

$$= P((X_1, \ldots, X_n) \in C_n \cap (-\infty, x_1] \times \cdots \times (-\infty, x_n])$$

$$= \int_{-\infty}^{x_1} \cdots \int_{-\infty}^{x_n} \prod_{i=1}^{n} f(y_i) \, I_{\{y_1 < \cdots < y_n\}} \, dy_n \cdots dy_1.$$

Therefore and since there are $n!$ elements in Π_n,

$$P\left(X_{(1)} \leq x_1, \ldots, X_{(n)} \leq x_n\right)$$

$$= \int_{-\infty}^{x_1} \cdots \int_{-\infty}^{x_n} n! \prod_{i=1}^{n} f(y_i) \, I_{\{y_1 < \cdots < y_n\}} \, dy_n \cdots dy_1. \qquad (2.1.14)$$

By Remark 2.1.10 about the support of $(X_{(1)}, \ldots, X_{(n)})$ and by virtue of the Radon-Nikodym theorem, we can read off the density of $(X_{(1)}, \ldots, X_{(n)})$ as the integrand in (2.1.14). Indeed, the Radon-Nikodym theorem ensures that the integrand is the a.e. unique probability density of $(X_{(1)}, \ldots, X_{(n)})$.[11] $\square$

We are now ready to formulate one of the main results of this course.

Theorem 2.1.11 (Order statistics property of the Poisson process)
Consider the Poisson process $N = (N(t))_{t \geq 0}$ with continuous a.e. positive intensity function λ and arrival times $0 < T_1 < T_2 < \cdots$ a.s. Then the conditional distribution of $(T_1, \ldots, T_n)$ given $\{N(t) = n\}$ is the distribution of the ordered sample $(X_{(1)}, \ldots, X_{(n)})$ of an iid sample $X_1, \ldots, X_n$ with common density $\lambda(x)/\mu(t)$, $0 < x \leq t$:

$$(T_1, \ldots, T_n \mid N(t) = n) \stackrel{d}{=} (X_{(1)}, \ldots, X_{(n)}).$$

In other words, the left-hand vector has conditional density

$$f_{T_1, \ldots T_n}(x_1, \ldots, x_n \mid N(t) = n) = \frac{n!}{(\mu(t))^n} \prod_{i=1}^{n} \lambda(x_i), \qquad (2.1.15)$$

$$0 < x_1 < \cdots < x_n < t.$$

Proof. We show that the limit

$$\lim_{h_i \downarrow 0, \, i=1,\ldots,n} \frac{P(T_1 \in (x_1, x_1 + h_1], \ldots, T_n \in (x_n, x_n + h_n] \mid N(t) = n)}{h_1 \cdots h_n}$$

$$(2.1.16)$$

exists and is a continuous function of the x_i's. A similar argument (which we omit) proves the analogous statement for the intervals $(x_i - h_i, x_i]$ with the same limit function. The limit can be interpreted as a density for the conditional probability distribution of $(T_1, \ldots, T_n)$, given $\{N(t) = n\}$.

Since $0 < x_1 < \cdots < x_n < t$ we can choose the h_i's so small that the intervals $(x_i, x_i + h_i] \subset [0, t]$, $i = 1, \ldots, n$, become disjoint. Then the following identity is immediate:

$$\{T_1 \in (x_1, x_1 + h_1], \ldots, T_n \in (x_n, x_n + h_n], N(t) = n\}$$

$$= \{N(0, x_1] = 0, N(x_1, x_1 + h_1] = 1, N(x_1 + h_1, x_2] = 0,$$

$$N(x_2, x_2 + h_2] = 1, \ldots, N(x_{n-1} + h_{n-1}, x_n] = 0,$$

$$N(x_n, x_n + h_n] = 1, N(x_n + h_n, t] = 0\}.$$

---

[11] Relation (2.1.14) means that for all rectangles $R = (-\infty, x_1] \times \cdots \times (-\infty, x_n]$ with $0 \leq x_1 < \cdots < x_n$ and for $\mathbf{X}_n = (X_{(1)}, \ldots, X_{(n)})$, $P(\mathbf{X}_n \in R) = \int_R f_{\mathbf{X}_n}(\mathbf{x}) \, d\mathbf{x}$. By the particular form of the support of $\mathbf{X}_n$, the latter relation remains valid for any rectangles in $\mathbb{R}^n$. An extension argument (cf. Billingsley [13]) ensures that the distribution of $\mathbf{X}_n$ is absolutely continuous with respect to Lebesgue measure with a density which coincides with $f_{\mathbf{X}_n}$ on the rectangles. The Radon-Nikodym theorem ensures the a.e. uniqueness of $f_{\mathbf{X}_n}$.

Taking probabilities on both sides and exploiting the independent increments of the Poisson process N, we obtain

$$P(T_1 \in (x_1, x_1 + h_1], \ldots, T_n \in (x_n, x_n + h_n], N(t) = n)$$

$$= P(N(0, x_1] = 0) \, P(N(x_1, x_1 + h_1] = 1) \, P(N(x_1 + h_1, x_2] = 0)$$

$$P(N(x_2, x_2 + h_2] = 1) \cdots P(N(x_{n-1} + h_{n-1}, x_n] = 0)$$

$$P(N(x_n, x_n + h_n] = 1) \, P(N(x_n + h_n, t] = 0)$$

$$= e^{-\mu(x_1)} \left[\mu(x_1, x_1 + h_1] e^{-\mu(x_1, x_1 + h_1]} \right] e^{-\mu(x_1 + h_1, x_2]}$$

$$\left[\mu(x_2, x_2 + h_2] e^{-\mu(x_2, x_2 + h_2]} \right] \cdots e^{-\mu(x_{n-1} + h_{n-1}, x_n]}$$

$$\left[\mu(x_n, x_n + h_n] e^{-\mu(x_n, x_n + h_n]} \right] e^{-\mu(x_n + h_n, t]}$$

$$= e^{-\mu(t)} \, \mu(x_1, x_1 + h_1] \cdots \mu(x_n, x_n + h_n].$$

Dividing by $P(N(t) = n) = e^{-\mu(t)}(\mu(t))^n/n!$ and $h_1 \cdots h_n$, we obtain the scaled conditional probability

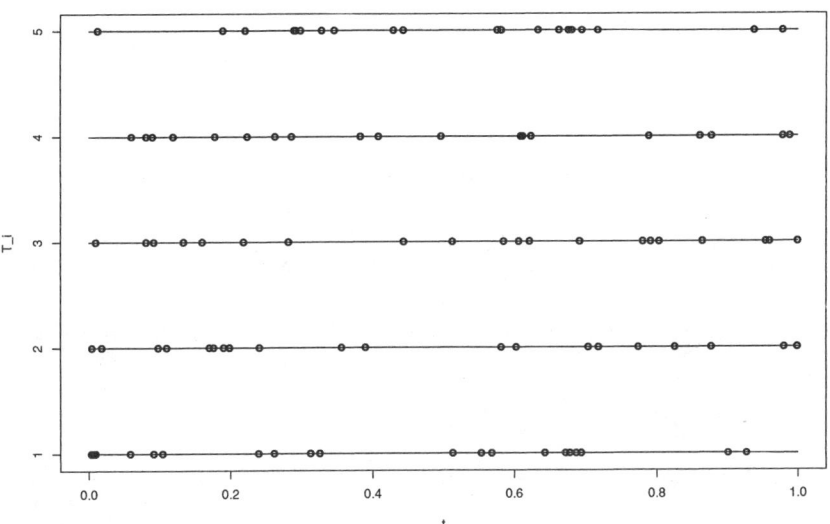

Figure 2.1.12 *Five realizations of the arrival times T_i of a standard homogeneous Poisson process conditioned to have 20 arrivals in $[0, 1]$. The arrivals in each row can be interpreted as the ordered sample of an iid $U(0, 1)$ sequence.*

$$\frac{P(T_1 \in (x_1, x_1 + h_1], \ldots, T_n \in (x_n, x_n + h_n] \mid N(t) = n)}{h_1 \cdots h_n}$$

$$= \frac{n!}{(\mu(t))^n} \frac{\mu(x_1, x_1 + h_1]}{h_1} \cdots \frac{\mu(x_n, x_n + h_n]}{h_n}$$

$$\to \frac{n!}{(\mu(t))^n} \lambda(x_1) \cdots \lambda(x_n), \quad \text{as } h_i \downarrow 0, \quad i = 1, \ldots, n.$$

Keeping in mind (2.1.16), this is the desired relation (2.1.15). In the last step we used the continuity of λ to show that $\mu'(x_i) = \lambda(x_i)$. □

Example 2.1.13 (Order statistics property of the homogeneous Poisson process)
Consider a homogeneous Poisson process with intensity $\lambda > 0$. Then Theorem 2.1.11 yields the joint conditional density of the arrival times T_i:

$$f_{T_1, \ldots, T_n}(x_1, \ldots, x_n \mid N(t) = n) = n! \, t^{-n}, \quad 0 < x_1 < \cdots < x_n < t.$$

A glance at Lemma 2.1.9 convinces one that this is the joint density of a uniform ordered sample $U_{(1)} < \cdots < U_{(n)}$ of iid $U(0, t)$ distributed $U_1, \ldots, U_n$. Thus, given there are n arrivals of a homogeneous Poisson process in the interval $[0, t]$, these arrivals constitute the points of a uniform ordered sample in $(0, t)$. In particular, this property is independent of the intensity λ! □

Example 2.1.14 (Symmetric function)
We consider a symmetric measurable function g on $\mathbb{R}^n$, i.e., for any permutation π of $\{1, \ldots, n\}$ we have

$$g(x_1, \ldots, x_n) = g(x_{\pi(1)}, \ldots, x_{\pi(n)}).$$

Such functions include products and sums:

$$g_s(x_1, \ldots, x_n) = \sum_{i=1}^n x_i, \quad g_p(x_1, \ldots, x_n) = \prod_{i=1}^n x_i.$$

Under the conditions of Theorem 2.1.11 and with the same notation, we conclude that

$$(g(T_1, \ldots, T_n) \mid N(t) = n) \overset{d}{=} g(X_{(1)}, \ldots, X_{(n)}) = g(X_1, \ldots, X_n).$$

For example, for any measurable function f on $\mathbb{R}$,

$$\left(\sum_{i=1}^n f(T_i) \; \middle| \; N(t) = n \right) \overset{d}{=} \sum_{i=1}^n f(X_{(i)}) = \sum_{i=1}^n f(X_i).$$

□

Example 2.1.15 (Shot noise)
This kind of stochastic process was used early on to model an electric current. Electrons arrive according to a homogeneous Poisson process N with rate λ at times T_i. An arriving electron produces an electric current whose time evolution of discharge is described as a deterministic function f with $f(t) = 0$ for $t < 0$. Shot noise describes the electric current at time t produced by all electrons arrived by time t as a superposition:

$$S(t) = \sum_{i=1}^{N(t)} f(t - T_i).$$

Typical choices for f are exponential functions $f(t) = e^{-\theta t} I_{[0,\infty)}(t)$, $\theta > 0$. An extension of classical shot noise processes with various applications is the process

$$S(t) = \sum_{i=1}^{N(t)} X_i f(t - T_i), \quad t \geq 0, \tag{2.1.17}$$

where

- (X_i) is an iid sequence, independent of (T_i).
- f is a deterministic function with $f(t) = 0$ for $t < 0$.

For example, if we assume that the X_i's are positive random variables, $S(t)$ is a generalization of the Cramér-Lundberg model, see Example 2.1.3. Indeed, choose $f = I_{[0,\infty)}$, then the shot noise process (2.1.17) is the total claim amount in the Cramér-Lundberg model. In an insurance context, f can also describe delay in claim settlement or some discount factor.

Delay in claim settlement is for example described by a function f satisfying

- $f(t) = 0$ for $t < 0$,
- $f(t)$ is non-decreasing,
- $\lim_{t \to \infty} f(t) = 1$.

In contrast to the Cramér-Lundberg model, where the claim size X_i is paid off at the time T_i when it occurs, a more general payoff function $f(t)$ allows one to delay the payment, and the speed at which this happens depends on the growth of the function f. Delay in claim settlement is advantageous from the point of view of the insurer. In the meantime the amount of money which was not paid for covering the claim could be invested and would perhaps bring some extra gain.

Suppose the amount Y_i is invested at time T_i in a riskless asset (savings account) with constant interest rate $r > 0$, (Y_i) is an iid sequence of positive random variables and the sequences (Y_i) and (T_i) are independent. Continuous compounding yields the amount $\exp\{r(t - T_i)\} Y_i$ at time $t > T_i$. For iid amounts Y_i which are invested at the arrival times T_i of a homogeneous Poisson process, the total value of all investments at time t is given by

$$S_1(t) = \sum_{i=1}^{N(t)} e^{r(t-T_i)} Y_i, \quad t \geq 0.$$

This is another shot noise process.

Alternatively, one may be interested in the present value of payments Y_i made at times T_i in the future. Then the present value with respect to the time frame $[0, t]$ is given as the *discounted sum*

$$S_2(t) = \sum_{i=1}^{N(t)} e^{-r(t-T_i)} Y_i, \quad t \geq 0.$$

A visualization of the sample paths of the processes S_1 and S_2 can be found in Figure 2.1.17. □

The distributional properties of a shot noise process can be treated in the framework of the following general result.

Proposition 2.1.16 *Let (X_i) be an iid sequence, independent of the sequence (T_i) of arrival times of a homogeneous Poisson process N with intensity λ. Then for any measurable function $g : \mathbb{R}^2 \to \mathbb{R}$ the following identity in distribution holds*

$$S(t) = \sum_{i=1}^{N(t)} g(T_i, X_i) \overset{d}{=} \sum_{i=1}^{N(t)} g(t U_i, X_i),$$

where (U_i) is an iid $U(0,1)$ sequence, independent of (X_i) and (T_i).

Proof. A conditioning argument together with the order statistics property of Theorem 2.1.11 yields that for $x \in \mathbb{R}$,

$$P\left(\sum_{i=1}^{N(t)} g(T_i, X_i) \leq x \,\middle|\, N(t) = n\right) = P\left(\sum_{i=1}^{n} g(t U_{(i)}, X_i) \leq x\right),$$

where $U_1, \ldots, U_n$ is an iid $U(0,1)$ sample, independent of (X_i) and (T_i), and $U_{(1)}, \ldots, U_{(n)}$ is the corresponding ordered sample. By the iid property of (X_i) and its independence of (U_i), we can permute the order of the X_i's arbitrarily without changing the distribution of $\sum_{i=1}^{n} g(t U_{(i)}, X_i)$:

$$P\left(\sum_{i=1}^{n} g(t U_{(i)}, X_i) \leq x\right) = E\left[P\left(\sum_{i=1}^{n} g(t U_{(i)}, X_i) \leq x \,\middle|\, U_1, \ldots, U_n\right)\right]$$

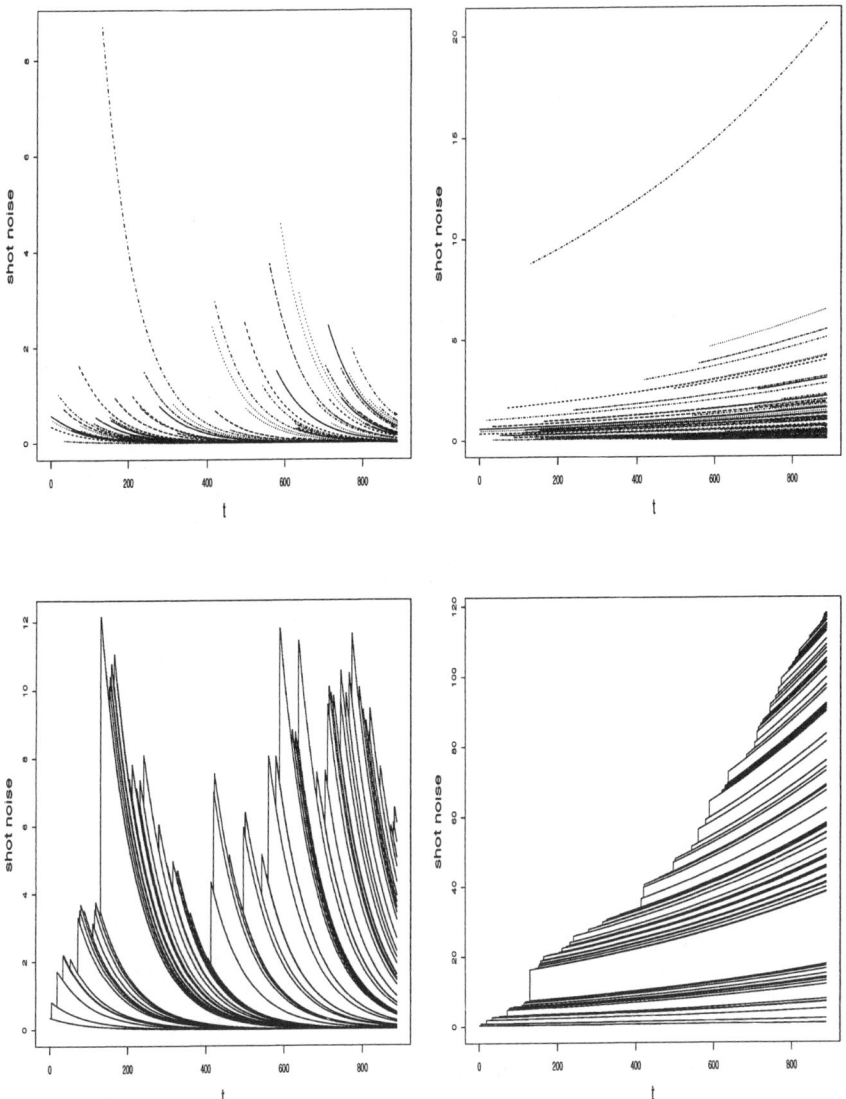

Figure 2.1.17 *Visualization of the paths of a shot noise process.* Top: 80 *paths of the processes* $Y_i \, \mathrm{e}^{\, r \, (t - T_i)}$, $t \geq T_i$, *where* (T_i) *are the point of a Poisson process with intensity* 0.1, (Y_i) *are iid standard exponential,* $r = -0.01$ *(left) and* $r = 0.001$ *(right).* Bottom: *The corresponding paths of the shot noise process* $S(t) = \sum_{T_i \leq t} Y_i \, \mathrm{e}^{\, r \, (t - T_i)}$ *presented as a superposition of the paths in the corresponding top graphs. The graphs show nicely how the interest rate r influences the aggregated value of future claims or payments Y_i. We refer to Example 2.1.15 for a more detailed description of these processes.*

$$= E\left[P\left(\sum_{i=1}^{n} g(t\,U_{(i)}, X_{\pi(i)}) \le x \,\Big|\, U_1,\ldots,U_n\right)\right], \qquad (2.1.18)$$

where π is any permutation of $\{1,\ldots,n\}$. In particular, we can choose π such that for given $U_1,\ldots,U_n$, $U_{(i)} = U_{\pi(i)}$, $i = 1,\ldots,n$.[12] Then (2.1.18) turns into

$$E\left[P\left(\sum_{i=1}^{n} g(t\,U_{\pi(i)}, X_{\pi(i)}) \le x \,\Big|\, U_1,\ldots,U_n\right)\right]$$

$$= E\left[P\left(\sum_{i=1}^{n} g(t\,U_i, X_i) \le x \,\Big|\, U_1,\ldots,U_n\right)\right]$$

$$= P\left(\sum_{i=1}^{n} g(t\,U_i, X_i) \le x\right) = P\left(\sum_{i=1}^{N(t)} g(t\,U_i, X_i) \le x \,\Big|\, N(t) = n\right).$$

Now it remains to take expectations:

$$P(S(t) \le x) = E[P(S(t) \le x \mid N(t))]$$

$$= \sum_{n=0}^{\infty} P(N(t) = n)\; P\left(\sum_{i=1}^{N(t)} g(T_i, X_i) \le x \,\Big|\, N(t) = n\right)$$

$$= \sum_{n=0}^{\infty} P(N(t) = n)\; P\left(\sum_{i=1}^{N(t)} g(t\,U_i, X_i) \le x \,\Big|\, N(t) = n\right)$$

$$= P\left(\sum_{i=1}^{N(t)} g(t\,U_i, X_i) \le x\right).$$

This proves the proposition.   □

---

[12] We give an argument to make this step in the proof more transparent. Since (U_i) and (X_i) are independent, it is possible to define $((U_i),(X_i))$ on the product space $\Omega_1 \times \Omega_2$ equipped with suitable σ-fields and probability measures, and such that (U_i) lives on Ω_1 and (X_i) on (Ω_2). While conditioning on $u_1 = U_1(\omega_1),\ldots,u_n = U_n(\omega_1)$, $\omega_1 \in \Omega_1$, choose the permutation $\pi = \pi(\omega_1)$ of $\{1,\ldots,n\}$ with $u_{\pi(1,\omega_1)} \le \cdots \le u_{\pi(n,\omega_1)}$, and then with probability 1,

$$P(\{\omega_2 : (X_1(\omega_2),\ldots,X_n(\omega_2)) \in A\}) =$$
$$P(\{\omega_2 : (X_{\pi(1,\omega_1)}(\omega_2),\ldots,X_{\pi(n,\omega_1)}(\omega_2))\} \in A \mid U_1(\omega_1) = u_1,\ldots,U_n(\omega_1) = u_n).$$

It is clear that Proposition 2.1.16 can be extended to the case when (T_i) is the arrival sequence of an inhomogeneous Poisson process. The interested reader is encouraged to go through the steps of the proof in this more general case.

Proposition 2.1.16 has a multitude of applications. We give one of them and consider more in the exercises.

Example 2.1.18 (Continuation of the shot noise Example 2.1.15)
In Example 2.1.15 we considered the stochastically discounted random sums

$$S(t) = \sum_{i=1}^{N(t)} e^{-r\,(t-T_i)}\, X_i\,. \tag{2.1.19}$$

According to Proposition 2.1.16, we have

$$S(t) \stackrel{d}{=} \sum_{i=1}^{N(t)} e^{-r\,(t-tU_i)}\, X_i \stackrel{d}{=} \sum_{i=1}^{N(t)} e^{-r\,t\,U_i}\, X_i\,, \tag{2.1.20}$$

where (X_i), (U_i) and N are mutually independent. Here we also used the fact that $(1 - U_i)$ and (U_i) have the same distribution. The structure of the random sum (2.1.19) is more complicated than the structure of the right-hand expression in (2.1.20) since in the latter sum the summands are independent of $N(t)$ and iid. For example, it is an easy matter to calculate the mean and variance of the expression on the right-hand side of (2.1.20) whereas it is a rather tedious procedure if one starts with (2.1.19). For example, we calculate

$$ES(t) = E\left(\sum_{i=1}^{N(t)} e^{-r\,t\,U_i}\, X_i\right) = E\left[E\left(\sum_{i=1}^{N(t)} e^{-r\,t\,U_i}\, X_i \,\middle|\, N(t)\right)\right]$$

$$= E\left[N(t)E\left(e^{-r\,t\,U_1}\, X_1\right)\right]$$

$$= EN(t)\, Ee^{-r\,t\,U_1}\, EX_1 = \lambda\,r^{-1}(1 - e^{-r\,t})\, EX_1\,.$$

Compare with the expectation in the Cramér-Lundberg model ($r = 0$): $ES(t) = \lambda\,t\,EX_1$. □

Comments

The order statistics property of a Poisson process can be generalized to Poisson processes with points in abstract spaces. We give an informal discussion of these processes in Section 2.1.8. In Exercise 20 on p. 58 we indicate how the "order statistics property" can be implemented, for example, in a Poisson process with points in the unit cube of $\mathbb{R}^d$.

2.1.7 A Discussion of the Arrival Times of the Danish Fire Insurance Data 1980-1990

In this section we want to illustrate the theoretical results of the Poisson process by means of the arrival process of a real-life data set: the *Danish fire insurance data* in the period from January 1, 1980, until December 31, 1990. The data were communicated to us by Mette Rytgaard and are available under www.ethz.ch/~mcneil. There is a total of $n = 2\,167$ observations. Here we focus on the arrival process. In Section 3.2, and in particular in Example 3.2.11, we study the corresponding claim sizes.

The arrival and the corresponding inter-arrival times are plotted in Figure 2.1.19. Together with the arrival times we show the straight line $\mu(t) = 1.85\,t$. The value $\widehat{\lambda} = n/T_n = 1/1.85$ is the maximum likelihood estimator of λ under the hypothesis that the inter-arrival times W_i are iid $\mathrm{Exp}(\lambda)$.

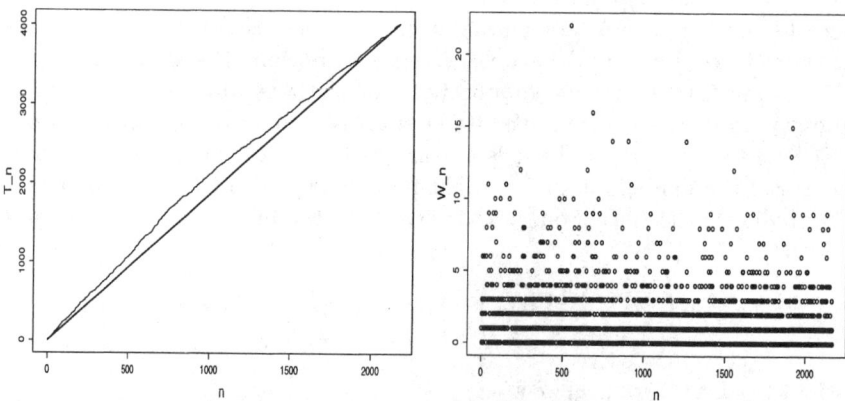

Figure 2.1.19 Left: *The arrival times of the Danish fire insurance data* 1980−1990. *The solid straight line has slope* 1.85 *which is estimated as the overall sample mean of the inter-arrival times. Since the graph of* (T_n) *lies above the straight line an inhomogeneous Poisson process is more appropriate for modeling the claim number in this portfolio.* Right: *The corresponding inter-arrival times. There is a total of* $n = 2\,167$ *observations.*

In Table 2.1.20 we summarize some basic statistics of the inter-arrival times for each year and for the whole period. Since the reciprocal of the annual sample mean is an estimator of the intensity, the table gives one the impression that there is a tendency for increasing intensity when time goes by. This phenomenon is supported by the left graph in Figure 2.1.21 where the annual mean inter-arrival times are visualized together with moving average estimates of the intensity function $\lambda(t)$. The estimate of the mean inter-arrival time at $t = i$ is defined as the moving average[13]

$$(\widehat{\lambda}(i))^{-1} = (2m + 1)^{-1} \sum_{j=\max(1,i-m)}^{\min(n,i+m)} W_j \quad \text{for } m = 50. \qquad (2.1.21)$$

The corresponding estimates for $\widehat{\lambda}(i)$ can be interpreted as estimates of the intensity function. There is a clear tendency for the intensity to increase over the last years. This tendency can also be seen in the right graph of Figure 2.1.21. Indeed, the boxplots[14] of this figure indicate that the distribution of the inter-arrival times of the claims is less spread towards the end of the 1980s and concentrated around the value 1 in contrast to 2 at the beginning of the 1980s. Moreover, the annual claim number increases.

year	1980	1981	1982	1983	1984	1985	1986	1987	1988	1989	1990	all
sample size	166	170	181	153	163	207	238	226	210	235	218	2 167
min	0	0	0	0	0	0	0	0	0	0	0	0
1st quartile	1	1	0.75	1	1	1	0	0	0	0	0	1
median	2	2	1	2	1.5	1	1	1	1	1	1	1
mean	2.19	2.15	1.99	2.37	2.25	1.76	1.53	1.62	1.73	1.55	1.68	1.85
$\widehat{\lambda}$ =1/mean	0.46	0.46	0.50	0.42	0.44	0.57	0.65	0.62	0.58	0.64	0.59	0.54
3rd quartile	3	3	3	3	3	2	2	2	3	2	2	3
max	11	12	10	22	16	14	14	9	12	15	9	22

Table 2.1.20 *Basic statistics for the Danish fire inter-arrival times data.*

Since we have gained statistical evidence that the intensity function of the Danish fire insurance data is not constant over 11 years, we assume in Figure 2.1.22 that the arrivals are modeled by an inhomogeneous Poisson process with continuous mean value function. We assume that the intensity is constant for every year, but it may change from year to year. Hence the mean value function $\mu(t)$ of the Poisson process is piecewise linear with possibly different slopes in different years; see the top left graph in Figure 2.1.22. We

---

[13] Moving average estimates such as (2.1.21) are proposed in time series analysis in order to estimate a deterministic trend which perturbs a stationary time series. We refer to Brockwell and Davis [16] and Priestley [63] for some theory and properties of the estimator $(\widehat{\lambda}(i))^{-1}$ and related estimates. More sophisticated estimators can be obtained by using kernel curve estimators in the regression model $W_i = (\lambda(i))^{-1} + \varepsilon_i$ for some smooth deterministic function λ and iid or weakly dependent stationary noise (ε_i). We refer to Fan and Gijbels [31] and Gasser et al. [33] for some standard theory of kernel curve estimation; see also Müller and Stadtmüller [59].

[14] The boxplot of a data set is a means to visualize the empirical distribution of the data. The middle part of the plot (box) indicates the median $x_{0.50}$, the 25% and 75% quantiles ($x_{0.25}$ and $x_{0.75}$) of the data. The "whiskers" of the data are the lines $x_{0.50} \pm 1.5\,(x_{0.75} - x_{0.25})$. Values outside the whiskers ("outliers") are plotted as points.

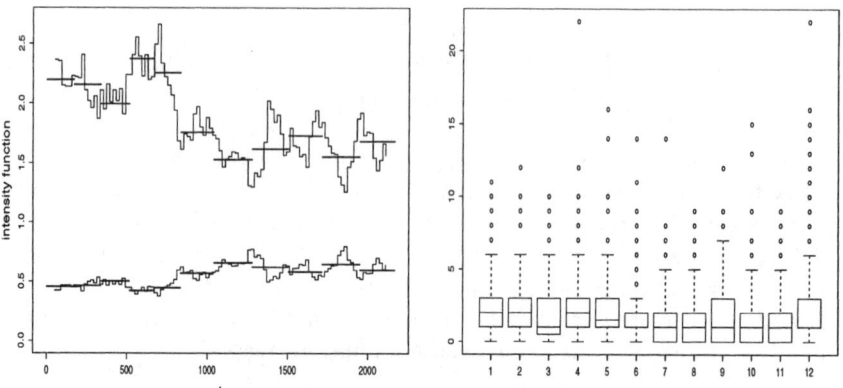

Figure 2.1.21 Left, upper graph: *The piecewise constant function represents the annual expected inter-arrival time between* 1980 *and* 1990. *The length of each constant piece is the claim number in the corresponding year. The annual estimates are supplemented by a moving average estimate* $(\widehat{\lambda}(i))^{-1}$ *defined in* (2.1.21). *Left, lower graph: The reciprocals of the values of the upper graph which can be interpreted as estimates of the Poisson intensity. There is a clear tendency for the intensity to increase over the last years.* Right: *Boxplots for the annual samples of the inter-arrival times* (No 1-11) *and the sample over* 11 *years* (No 12).

choose the estimated intensities presented in Table 2.1.20 and in the left graph of Figure 2.1.21. We transform the arrivals T_n into $\mu(T_n)$. According to the theory in Section 2.1.3, one can interpret the points $\mu(T_n)$ as arrivals of a standard homogeneous Poisson process. This is nicely illustrated in the top right graph of Figure 2.1.22, where the sequence $(\mu(T_n))$ is plotted against n. The graph is very close to a straight line, in contrast to the left graph in Figure 2.1.19, where one can clearly see the deviations of the arrivals T_n from a straight line.

In the left middle graph we consider the histogram of the time changed inter-arrival times $\mu(T_n)$. According to the theory in Section 2.1.6, the arrival times of a homogeneous Poisson can be interpreted as a uniform sample on any fixed interval, conditionally on the claim number in this interval. The histogram resembles the histogram of a uniform sample in contrast to the middle right graph, where the histogram of the Danish fire arrival times is presented. However, the left histogram is not perfect either. This is due to the fact that the data T_n are integers, hence the values $\mu(T_n)$ live on a particular discrete set.

The left bottom graph shows a moving average estimate of the intensity function of the arrivals $\mu(T_n)$. Although the function is close to 1 the estimates fluctuate wildly around 1. This is an indication that the process might not be Poisson and that other models for the arrival process could be more appropriate; see for example Section 2.2. The deviation of the distribution of the inter-arrival time $\mu(T_n) - \mu(T_{n-1})$, which according to the theory should be iid standard exponential, can also be seen in the right bottom graph in Figure 2.1.22, where a QQ-plot[15] of these data against the standard exponential distribution is shown. The QQ-plot curves down at the right. This is a clear indication of a right tail of the underlying distribution which is heavier than the tail of the exponential distribution. These observations raise the question as to whether the Poisson process is a suitable model for the whole period of 11 years of claim arrivals.

A homogeneous Poisson process is a suitable model for the arrivals of the Danish fire insurance data for shorter periods of time such as one year. This is illustrated in Figure 2.1.23 for the 166 arrivals in the period January 1 - December 31, 1980.

As a matter of fact, the data show a clear seasonal component. This can be seen in Figure 2.1.24, where a histogram of all arrivals modulo 366 is given. Hence one receives a distribution on the integers between 1 and 366. Notice for example the peak around day 120 which corresponds to fires in April-May. There is also more activity in summer than in early spring and late fall, and one observes more fires in December and January with the exception of the last week of the year.

2.1.8 An Informal Discussion of Transformed and Generalized Poisson Processes

Consider a Poisson process N with claim arrival times T_i on $[0, \infty)$ and mean value function μ, independent of the iid positive claim sizes X_i with distribution function F. In this section we want to learn about a procedure which allows one to merge the Poisson claim arrival times T_i and the iid claim sizes X_i in one Poisson process with points in $\mathbb{R}^2$.

Define the counting process

$$M(a,b) = \#\{i \geq 1 : X_i \leq a, T_i \leq b\} = \sum_{i=1}^{N(b)} I_{(0,a]}(X_i), \quad a, b \geq 0.$$

We want to determine the distribution of $M(a,b)$. For this reason, recall the characteristic function[16] of a Poisson random variable $M \sim \text{Pois}(\gamma)$:

---

[15] The reader who is unfamiliar with QQ-plots is referred to Section 3.2.1.

[16] In what follows we work with characteristic functions because this notion is defined for all distributions on $\mathbb{R}$. Alternatively, we could replace the characteristic functions by moment generating functions. However, the moment generating function of a random variable is well-defined only if this random variable has certain finite exponential moments. This would restrict the class of distributions we consider.

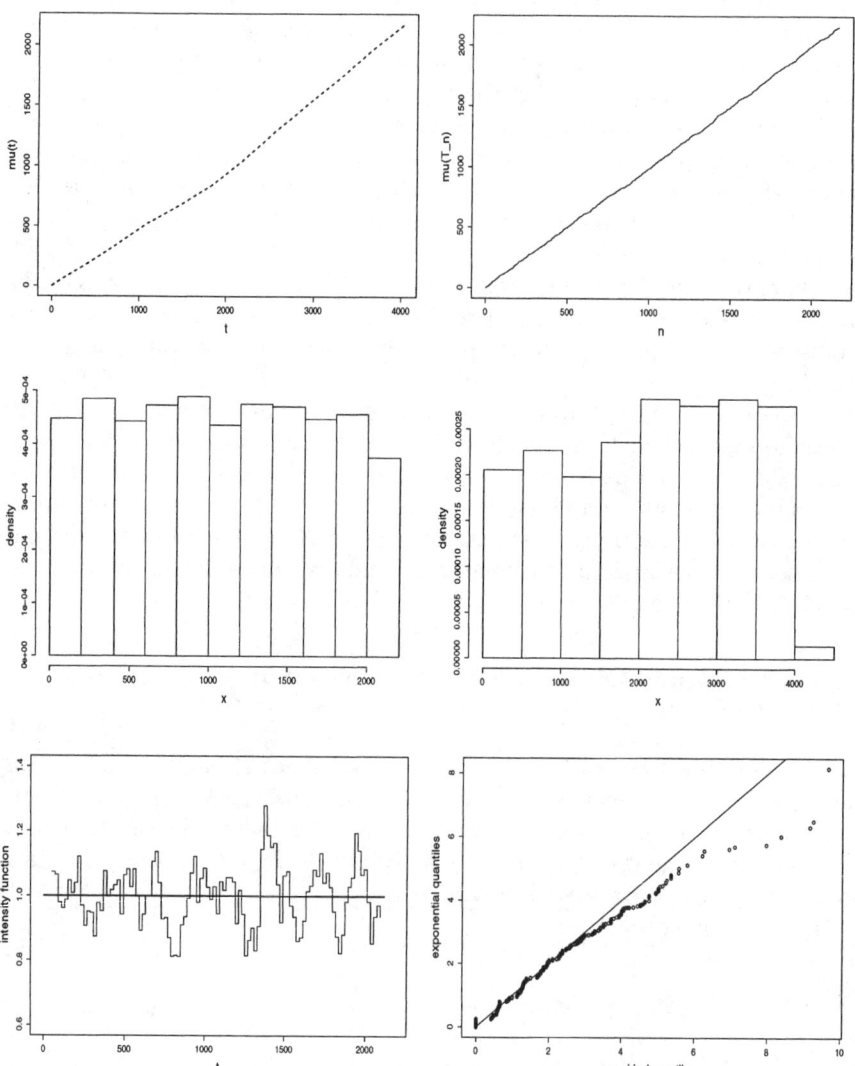

Figure 2.1.22 Top left: *The estimated mean value function $\mu(t)$ of the Danish fire insurance arrivals. The function is piecewise linear. The slopes are the estimated intensities from Table 2.1.20.* Top right: *The transformed arrivals $\mu(T_n)$. Compare with Figure 2.1.19. The histogram of the values $\mu(T_n)$ (middle left) resembles a uniform density, whereas the histogram of the T_n's shows clear deviations from it (middle right).* Bottom left: *Moving average estimate of the intensity function corresponding to the transformed sequence $(\mu(T_n))$. The estimates fluctuate around the value 1.* Bottom right: *QQ-plot of the values $\mu(T_n) - \mu(T_{n-1})$ against the standard exponential distribution. The plot curves down at the right end indicating that the values come from a distribution with tails heavier than exponential.*

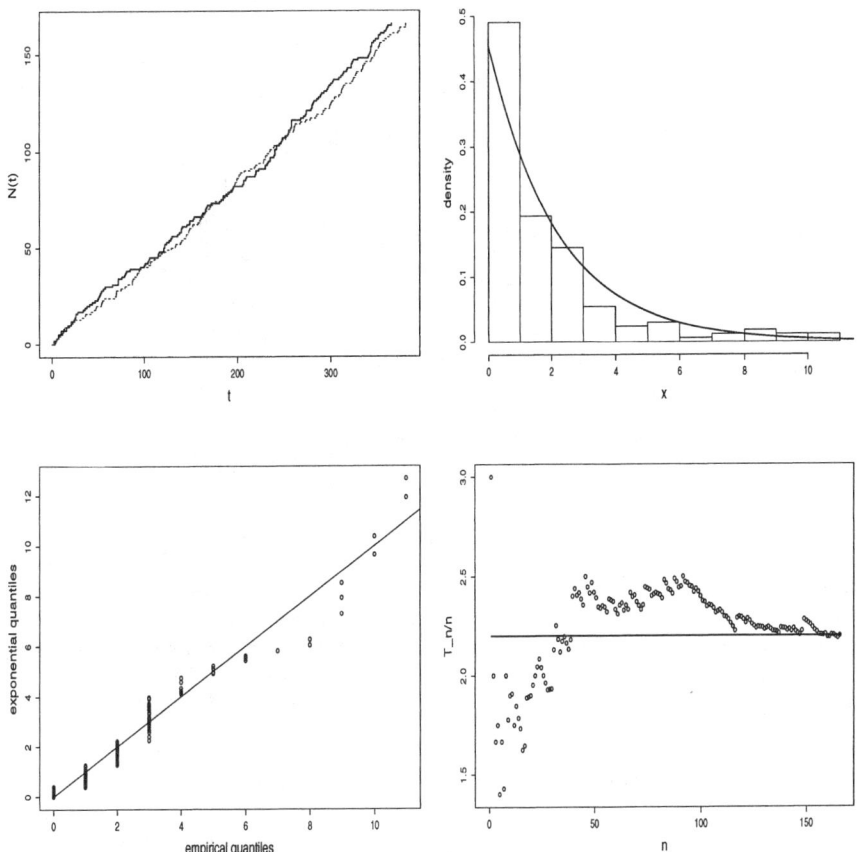

Figure 2.1.23 *The Danish fire insurance arrivals from January* 1, 1980, *until December* 31, 1980. *The inter-arrival times have sample mean* $\widehat{\lambda}^{-1} = 2.19$. Top left: *The renewal process* $N(t)$ *generated by the arrivals (solid boldface curve). For comparison, one sample path of a homogeneous Poisson process with intensity* $\lambda = (2.19)^{-1}$ *is drawn.* Top right: *The histogram of the inter-arrival times. For comparison, the density of the* $\mathrm{Exp}(\lambda)$ *distribution is drawn.* Bottom left: *QQ-plot for the inter-arrival sample against the quantiles of the* $\mathrm{Exp}(\lambda)$ *distribution. The fit of the data by an exponential* $\mathrm{Exp}(\lambda)$ *is not unreasonable. However, the QQ-plot indicates a clear difference to exponential inter-arrival times: the data come from an integer-valued distribution. This deficiency could be overcome if one knew the exact claim times.* Bottom right: *The ratio* T_n/n *as a function of time. The values cluster around* $\widehat{\lambda}^{-1} = 2.19$ *which is indicated by the constant line. For a homogeneous Poisson process,* $T_n/n \overset{a.s.}{\to} \lambda^{-1}$ *by virtue of the strong law of large numbers. For an iid* $\mathrm{Exp}(\lambda)$ *sample* $W_1, \ldots, W_n$, $\widehat{\lambda} = n/T_n$ *is the maximum likelihood estimator of* λ. *If one accepts the hypothesis that the arrivals in* 1980 *come from a homogeneous Poisson process with intensity* $\lambda = (2.19)^{-1}$, *one would have an expected inter-arrival time of* 2.19, *i.e., roughly every second day a claim occurs.*

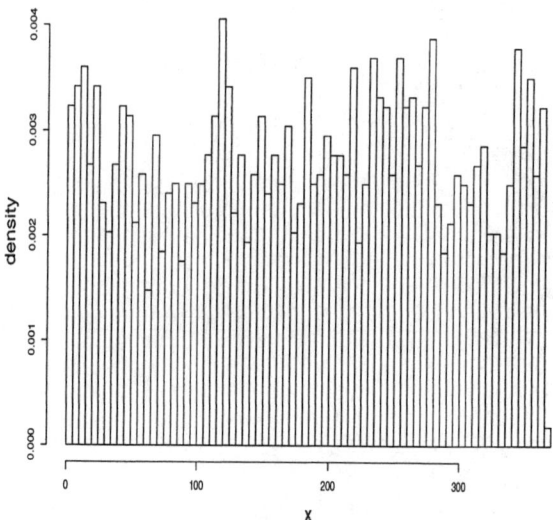

Figure 2.1.24 *Histogram of all arrival times of the Danish fire insurance claims considered as a distribution on the integers between 1 and 366. The bars of the histogram correspond to the weeks of the year. There is a clear indication of seasonality in the data.*

$$Ee^{itM} = \sum_{n=0}^{\infty} e^{itn} P(M = n) = \sum_{n=0}^{\infty} e^{itn} e^{-\gamma} \frac{\gamma^n}{n!} = e^{-\gamma(1-e^{it})}, \quad t \in \mathbb{R}.$$

$$(2.1.22)$$

We know that the characteristic function of a random variable M determines its distribution and vice versa. Therefore we calculate the characteristic function of $M(a,b)$. A similar argument as the one leading to (2.1.22) yields

$$Ee^{itM(a,b)} = E\left[E\exp\left\{it\sum_{j=1}^{N(b)} I_{(0,a]}(X_j)\right\}\middle| N(b)\right]$$

$$= E\left[\left(E\exp\left\{it\, I_{(0,a]}(X_1)\right\}\right)^{N(b)}\right]$$

$$= E\left(\left[1 - F(a) + F(a)\,e^{it}\right]^{N(b)}\right)$$

$$= e^{-\mu(b)\,F(a)\,(1-e^{it})}. \qquad (2.1.23)$$

We conclude from (2.1.22) and (2.1.23) that $M(a,b) \sim \text{Pois}(F(a)\,\mu(b))$. Using similar characteristic function arguments, one can show that

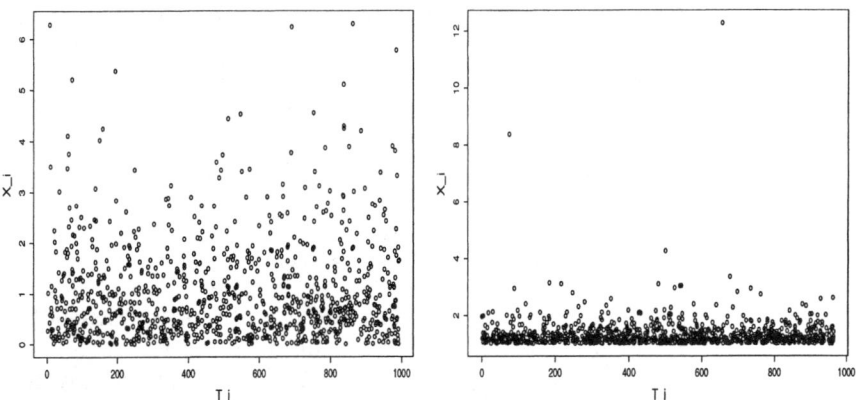

Figure 2.1.25 $1\,000$ *points* (T_i, X_i) *of a two-dimensional Poisson process, where* (T_i) *is the sequence of the the arrival times of a homogeneous Poisson process with intensity 1 and* (X_i) *is a sequence of iid claim sizes, independent of* (T_i). *Left: Standard exponential claim sizes. Right: Pareto distributed claim sizes with* $P(X_i > x) = x^{-4}$, $x \geq 1$. *Notice the difference in scale of the claim sizes!*

- The increments

$$M((x, x+h] \times (t, t+s])$$

$$= \#\{i \geq 1 : (X_i, T_i) \in (x, x+h] \times (t, t+s]\}, \quad x, t \geq 0, \; h, s > 0,$$

  are $\mathrm{Pois}(F(x, x+h] \, \mu(t, t+s])$ distributed.
- For disjoint intervals $\Delta_i = (x_i, x_i + h_i] \times (t_i, t_i + s_i]$, $i = 1, \ldots, n$, the increments $M(\Delta_i)$, $i = 1, \ldots, n$, are independent.

From measure theory, we know that the quantities $F(x, x+h] \, \mu(t, t+s]$ determine the product measure $\gamma = F \times \mu$ on the Borel σ-field of $[0, \infty)^2$, where F denotes the distribution function as well as the distribution of X_i and μ is the measure generated by the values $\mu(a, b]$, $0 \leq a < b < \infty$. This is a consequence of the extension theorem for measures; cf. Billingsley [13]. In the case of a homogeneous Poisson process, $\mu = \lambda\,\mathrm{Leb}$, where Leb denotes Lebesgue measure on $[0, \infty)$.

In analogy to the extension theorem for deterministic measures, one can find an extension M of the random counting variables $M(\Delta)$, $\Delta = (x, x+h] \times (t, t+s]$, such that for any Borel set[17] $A \subset [0, \infty)^2$,

$$M(A) = \#\{i \geq 1 : (X_i, T_i) \in A\} \sim \mathrm{Pois}(\gamma(A)),$$

and for disjoint Borel sets $A_1, \ldots, A_n \subset [0, \infty)^2$, $M(A_1), \ldots, M(A_n)$ are independent. We call $\gamma = F \times \mu$ the *mean measure* of M, and M is called a

---

[17] For A with mean measure $\gamma(A) = \infty$, we write $M(A) = \infty$.

Poisson process or a *Poisson random measure with mean measure* γ, denoted $M \sim \mathrm{PRM}(\gamma)$. Notice that M is indeed a random counting measure on the Borel σ-field of $[0, \infty)^2$.

The embedding of the claim arrival times and the claim sizes in a Poisson process with two-dimensional points gives one a precise answer as to how many claim sizes of a given magnitude occur in a fixed time interval. For example, the number of claims exceeding a high threshold u, say, in the period $(a, b]$ of time is given by

$$M((u, \infty) \times (a, b]) = \#\{i \geq 1 : X_i > u, T_i \in (a, b]\}.$$

This is a $\mathrm{Pois}((1 - F(u))\, \mu(a, b])$ distributed random variable. It is independent of the number of claims below the threshold u occurring in the same time interval. Indeed, the sets $(u, \infty) \times (a, b]$ and $[0, u] \times (a, b]$ are disjoint and therefore $M((u, \infty) \times (a, b])$ and $M([0, u] \times (a, b])$ are independent Poisson distributed random variables.

In the previous sections[18] we used various transformations of the arrival times T_i of a Poisson process N on $[0, \infty)$ with mean measure ν, say, to derive other Poisson processes on the interval $[0, \infty)$. The restriction of processes to $[0, \infty)$ can be relaxed. Consider a measurable set $E \subset \mathbb{R}$ and equip E with the σ-field $\mathcal{E}$ of the Borel sets. Then

$$N(A) = \#\{i \geq 1 : T_i \in A\}, \quad A \in \mathcal{E},$$

defines a *random measure* on the measurable space $(E, \mathcal{E})$. Indeed, $N(A) = N(A, \omega)$ depends on $\omega \in \Omega$ and for fixed ω, $N(\cdot, \omega)$ is a counting measure on $\mathcal{E}$. The set E is called the *state space* of the random measure N. It is again called a *Poisson random measure* or *Poisson process* with mean measure ν restricted to E since one can show that $N(A) \sim \mathrm{Pois}(\nu(A))$ for $A \in \mathcal{E}$, and $N(A_i)$, $i = 1, \ldots, n$, are mutually independent for disjoint $A_i \in \mathcal{E}$. The notion of Poisson random measure is very general and can be extended to abstract state spaces E. At the beginning of the section we considered a particular example in $E = [0, \infty)^2$. The Poisson processes we considered in the previous sections are examples of Poisson processes with state space $E = [0, \infty)$.

One of the strengths of this general notion of Poisson process is the fact that *Poisson random measures remain Poisson random measures under measurable transformations*. Indeed, let $\psi : E \to \widetilde{E}$ be such a transformation and $\widetilde{E}$ be equipped with the σ-field $\widetilde{\mathcal{E}}$. Assume N is $\mathrm{PRM}(\nu)$ on E with points T_i. Then the points $\psi(T_i)$ are in $\widetilde{E}$ and, for $A \in \widetilde{\mathcal{E}}$,

$$N_\psi(A) = \#\{i \geq 1 : \psi(T_i) \in A\} = \#\{i \geq 1 : T_i \in \psi^{-1}(A)\} = N(\psi^{-1}(A)),$$

where $\psi^{-1}(A) = \{x \in E : \psi(x) \in A\}$ denotes the inverse image of A which belongs to $\mathcal{E}$ since ψ is measurable. Then we also have that $N_\psi(A) \sim \mathrm{Pois}(\nu(\psi^{-1}(A)))$ since $EN_\psi(A) = EN(\psi^{-1}(A)) = \nu(\psi^{-1}(A))$. Moreover,

---

[18] See, for example, Section 2.1.3.

since disjointness of $A_1, \ldots, A_n$ in $\widetilde{\mathcal{E}}$ implies disjointness of $\psi^{-1}(A_1), \ldots,$
$\psi^{-1}(A_n)$ in $\mathcal{E}$, it follows that $N_\psi(A_1), \ldots, N_\psi(A_n)$ are independent, by the
corresponding property of the PRM N. We conclude that $N_\psi \sim \mathrm{PRM}(\nu(\psi^{-1}))$.

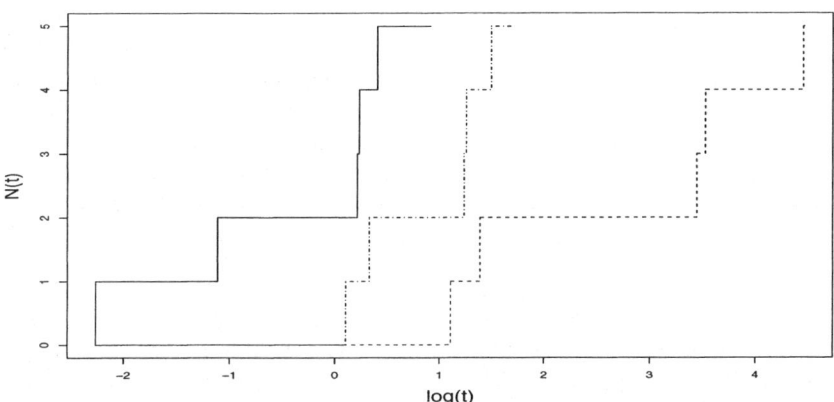

log(t)

Figure 2.1.26 *Sample paths of the Poisson processes with arrival times* $\exp\{T_i\}$
(bottom dashed curve), T_i *(middle dashed curve) and* $\log T_i$ *(top solid curve). The*
T_i*'s are the arrival times of a standard homogeneous Poisson process. Time is on*
logarithmic scale in order to visualize the three paths in one graph.

Example 2.1.27 (Measurable transformations of Poisson processes remain
Poisson processes)
(1) Let $\widetilde{N}$ be a Poisson process on $[0, \infty)$ with mean value function $\widetilde{\mu}$ and
arrival times $0 < T_1 < T_2 < \cdots$. Consider the transformed process

$$N(t) = \#\{i \geq 1 : T_i\, I_{[a,b]}(T_i) \leq t\}, \quad t \geq 0,$$

for some interval $[a, b] \subset [0, \infty)$, where $\psi(x) = x\, I_{[a,b]}(x)$ is clearly measurable.
This construction implies that $N(A) = \#\{i \geq 1 : T_i I_{[a,b]}(T_i) \in A\} = 0$ for
$A \subset [a, b]^c$, the complement of $[a, b]$. Therefore it suffices to consider N on the
Borel sets of $[a, b]$. This defines a *Poisson process on* $[a, b]$ with mean value
function $\mu(t) = \widetilde{\mu}(t) - \widetilde{\mu}(a)$, $t \in [a, b]$.
(2) Consider a standard homogeneous Poisson process on $[0, \infty)$ with arrival
times $0 < T_1 < T_2 < \cdots$. We transform the arrival times with the measurable
function $\psi(x) = \log x$. Then the points $(\log T_i)$ constitute a Poisson process
N on $\mathbb{R}$. The Poisson measure of the interval $(a, b]$ for $a < b$ is given by

$$N(a, b] = \#\{i \geq 1 : \log(T_i) \in (a, b]\} = \#\{i \geq 1 : T_i \in (e^a, e^b]\}.$$

This is a $\mathrm{Pois}(e^b - e^a)$ distributed random variable, i.e., the mean measure
of the interval $(a, b]$ is given by $e^b - e^a$.

Alternatively, transform the arrival times T_i by the exponential function. The resulting Poisson process M is defined on $[1, \infty)$. The Poisson measure of the interval $(a, b] \subset [1, \infty)$ is given by

$$M(a, b] = \#\{i \geq 1 : e^{T_i} \in (a, b]\} = \#\{i \geq 1 : T_i \in (\log a, \log b]\}.$$

This is a Pois$(\log(b/a))$ distributed random variable, i.e., the mean measure of the interval $(a, b]$ is given by $\log(b/a)$. Notice that this Poisson process has the remarkable property that $M(ca, cb]$ for any $c \geq 1$ has the same Pois$(\log(b/a))$ distribution as $M(a, b]$. In particular, the expected number of points $\exp\{T_i\}$ falling into the interval $(ca, cb]$ is independent of the value $c \geq 1$. This is somewhat counterintuitive since the length of the interval $(ca, cb]$ can be arbitrarily large. However, the larger the value c the higher the threshold ca which prevents sufficiently many points $\exp\{T_i\}$ from falling into the interval $(ca, cb]$, and on average there are as many points in $(ca, cb]$ as in $(a, b]$.   □

Example 2.1.28 (Construction of transformed planar PRM)
Let (T_i) be the arrival sequence of a standard homogeneous Poisson process on $[0, \infty)$, independent of the iid sequence (X_i) with common distribution function F. Then the points (T_i, X_i) constitute a PRM(ν) N with state space $E = [0, \infty) \times \mathbb{R}$ and mean measure $\nu = \text{Leb} \times F$; see the discussion on p. 45.

After a measurable transformation $\psi : \mathbb{R}^2 \to \mathbb{R}^2$ the points $\psi(T_i, X_i)$ constitute a PRM N_ψ with state space $E_\psi = \{\psi(t, x) : (t, x) \in E\}$ and mean measure $\nu_\psi(A) = \nu(\psi^{-1}(A))$ for any Borel set $A \subset E_\psi$. We choose $\widetilde{\psi}(t, x) = t^{-1/\alpha}(\cos(2\pi x), \sin(2\pi x))$ for some $\alpha \neq 0$, i.e., the PRM $N_{\widetilde{\psi}}$ has points $\mathbf{Y}_i = T_i^{-1/\alpha}(\cos(2\pi X_i), \sin(2\pi X_i))$. In Figure 2.1.30 we visualize the points $\mathbf{Y}_i$ of the resulting PRM for different choices of α and distribution functions F of X_1.

Planar PRMs such as the ones described above are used, among others, in spatial statistics (see Cressie [24]) in order to describe the distribution of random configurations of points in the plane such as the distribution of minerals, locations of highly polluted spots or trees in a forest. The particular PRM $N_{\widetilde{\psi}}$ and its modifications are major models in multivariate extreme value theory. It describes the dependence of extremes in the plane and in space. In particular, it is suitable for modeling clustering behavior of points $\mathbf{Y}_i$ far away from the origin. See Resnick [64] for the theoretical background on multivariate extreme value theory and Mikosch [58] for a recent attempt to use $N_{\widetilde{\psi}}$ for modeling multivariate financial time series.   □

Example 2.1.29 (Modeling arrivals of Incurred But Not Reported (IBNR) claims)
In a portfolio, the claims are not reported at their arrival times T_i, but with a certain delay. This delay may be due to the fact that the policyholder is not aware of the claim and only realizes it later (for example, a damage in his/her house), or that the policyholder was injured in a car accident and did not have the opportunity to call his agent immediately, or the policyholder's

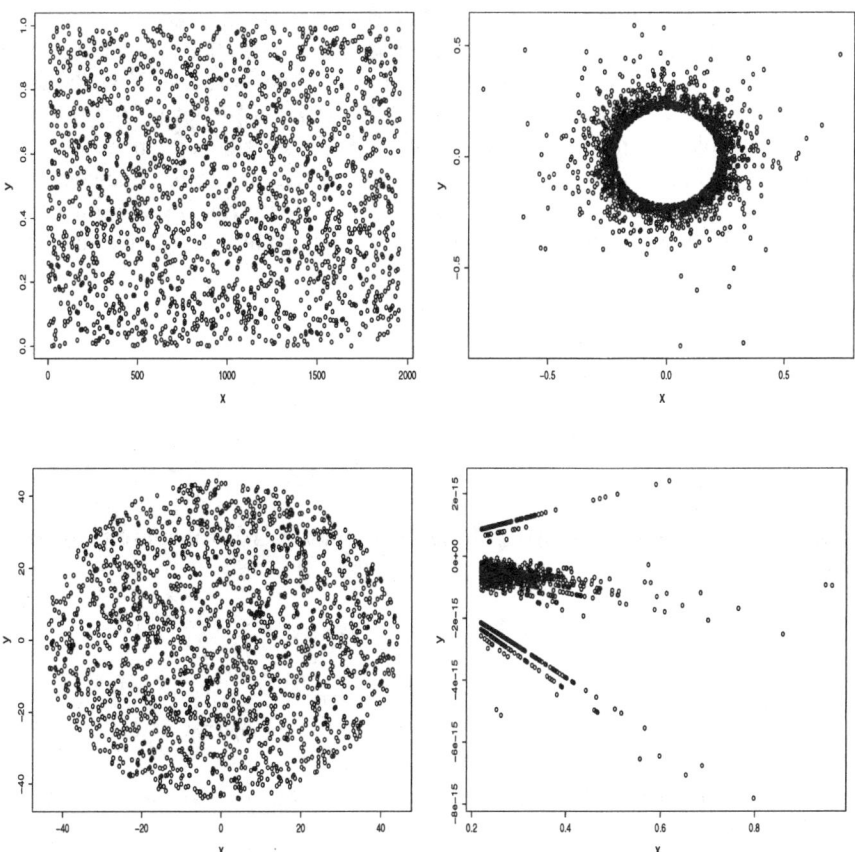

Figure 2.1.30 *Poisson random measures in the plane.*

Top left: 2 000 *points of a Poisson random measure with points* (T_i, X_i), *where* (T_i) *is the arrival sequence of a standard homogeneous Poisson process on* $[0, \infty)$, *independent of the iid sequence* (X_i) *with* $X_1 \sim \mathrm{U}(0, 1)$. *The PRM has mean measure* $\nu = \mathrm{Leb} \times \mathrm{Leb}$ *on* $[0, \infty) \times (0, 1)$.

After the measurable transformation $\widetilde{\psi}(t, x) = t^{-1/\alpha} (\cos(2\pi x), \sin(2\pi x))$ *for some* $\alpha \neq 0$ *the resulting PRM* $N_{\widetilde{\psi}}$ *has points* $\mathbf{Y}_i = T_i^{-1/\alpha}(\cos(2\pi X_i), \sin(2\pi X_i))$.

Top right: *The points of the process* $N_{\widetilde{\psi}}$ *for* $\alpha = 5$ *and iid* $\mathrm{U}(0, 1)$ *uniform* X_i's. *Notice that the spherical part* $(\cos(2\pi X_i), \sin(2\pi X_i))$ *of* $\mathbf{Y}_i$ *is uniformly distributed on the unit circle.*

Bottom left: *The points of the process* $N_{\widetilde{\psi}}$ *with* $\alpha = -5$ *and iid* $\mathrm{U}(0, 1)$ *uniform* X_i's.

Bottom right: *The points of the process* $N_{\widetilde{\psi}}$ *for* $\alpha = 5$ *with iid* $X_i \sim \mathrm{Pois}(10)$.

flat burnt down over Christmas, but the agent was on a skiing vacation in Switzerland and could not receive the report about the fire, etc.

We consider a simple model for the reporting times of IBNR claims: the arrival times T_i of the claims are modeled by a Poisson process N with mean value function μ and the delays in reporting by an iid sequence (V_i) of positive random variables with common distribution F. Then the sequence $(T_i + V_i)$ constitutes the reporting times of the claims to the insurance business. We assume that (V_i) and (T_i) are independent. Then the points (T_i, V_i) constitute a PRM(ν) with mean measure $\nu = \mu \times F$. By time t, $N(t)$ claims have occurred, but only

$$N_{\text{IBNR}}(t) = \sum_{i=1}^{N(t)} I_{[0,t]}(T_i + V_i) = \#\{i \geq 1 : T_i + V_i \leq t\}$$

have been reported. The mapping $\psi(t, v) = t + v$ is measurable. It transforms the points (T_i, V_i) of the PRM(ν) into the points $T_i + V_i$ of the PRM N_ψ with mean measure of a set A given by $\nu_\psi(A) = \nu(\psi^{-1}(A))$. In particular, $N_{\text{IBNR}}(s) = N_\psi([0, s])$ is Pois($\nu_\psi([0, s])$) distributed. We calculate the mean value

$$\nu_\psi([0, s]) = (\mu \times F)\{(t, v) : 0 \leq t + v \leq s\}$$

$$= \int_{t=0}^{s} \int_{v=0}^{s-t} dF(v)\, d\mu(t) = \int_{0}^{s} F(s - t)\, d\mu(t)\,.$$

If N is homogeneous Poisson with intensity $\lambda > 0$, $\mu = \lambda\,\text{Leb}$, and then

$$\nu_\psi([0, s]) = \lambda \int_{0}^{s} F(t)\, dt = \lambda\, s - \lambda \int_{0}^{s} \overline{F}(t)\, dt\,, \qquad (2.1.24)$$

where $\overline{F} = 1 - F$ is the tail of the distribution function F. The second term in (2.1.24) converges to the value $\lambda\, EV_1 = \lambda \int_{0}^{\infty} \overline{F}(t) dt$ as $s \to \infty$. The delayed claim numbers $N_{\text{IBNR}}(s)$ constitute an inhomogeneous Poisson process on $[0, \infty)$ whose mean value function differs from $EN(s) = \lambda s$ by the value $\lambda \int_{0}^{s} \overline{F}(t)\, dt$. If $EV_1 < \infty$ and $h > 0$ is fixed, the difference of the mean values of the increments $N(s, s+h]$ and $N_{\text{IBNR}}(s, s+h]$ is asymptotically negligible. $\square$

Comments

The Poisson process is one of the most important stochastic processes. For the abstract understanding of this process one would have to consider it as a *point process*, i.e., as a random counting measure. We have indicated in Section 2.1.8 how one has to approach this problem. As a matter of fact, various other counting processes such as the renewal process treated in Section 2.2 are

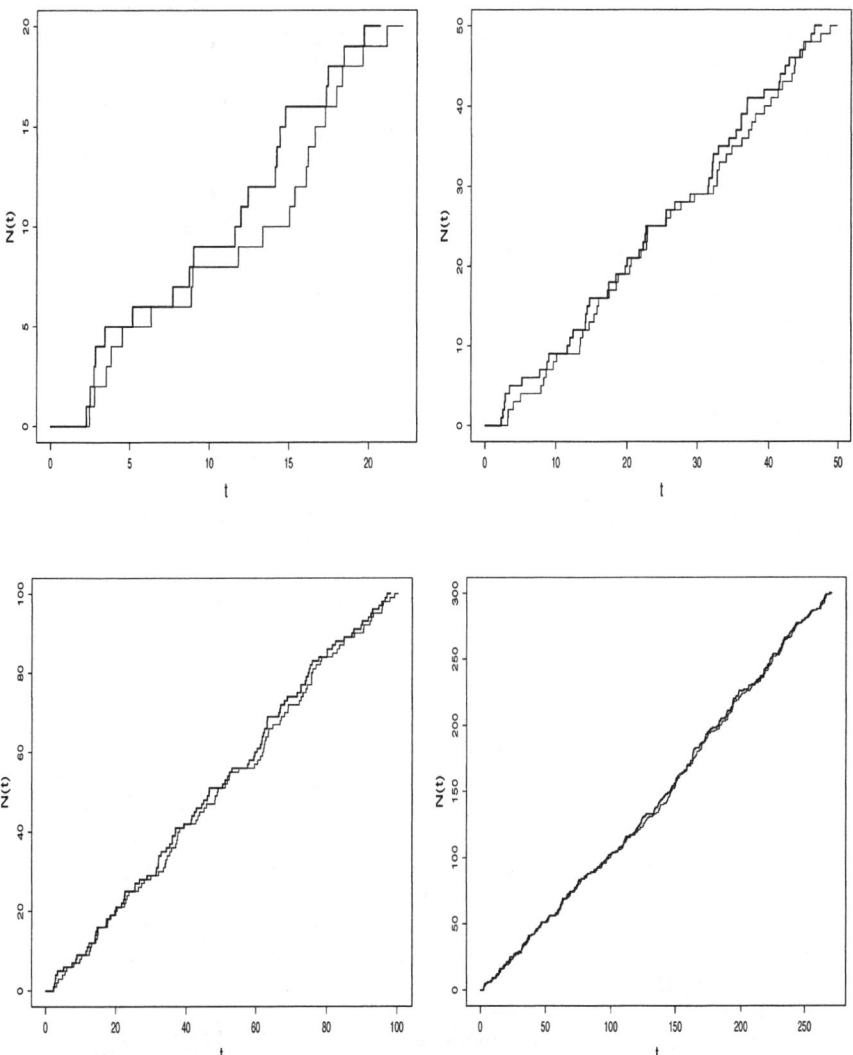

Figure 2.1.31 *Incurred But Not Reported claims. We visualize one sample of a standard homogeneous Poisson process with n arrivals T_i (top boldface graph) and the corresponding claim number process for the delayed process with arrivals $T_i + V_i$, where the V_i's are iid Pareto distributed with distribution $P(V_1 > x) = x^{-2}$, $x \geq 1$, independent of (T_i). Top: $n = 30$ (left) and $n = 50$ (right). Bottom: $n = 100$ (left) and $n = 300$ (right). As explained in Example 2.1.29, the sample paths of the claim number process differ from each other approximately by the constant value EV_1. For sufficiently large t, the difference is negligible compared to the expected claim number.*

approximated by suitable Poisson processes in the sense of convergence in distribution. Therefore the Poisson process with nice mathematical properties is also a good approximation to various real-life counting processes such as the claim number process in an insurance portfolio.

The treatment of general Poisson processes requires more stochastic process theory than available in this course. For a gentle introduction we refer to Embrechts et al. [29], Chapter 5; for a rigorous treatment at a moderate level, Resnick's [65] monograph or Kingman's book [50] are good references. Resnick's monograph [64] is a more advanced text on the Poisson process with various applications to extreme value theory. See also Daley and Vere-Jones [25] or Kallenberg [48] for some advanced treatments.

Exercises

Sections 2.1.1-2.1.2

(1) Let $N = (N(t))_{t \geq 0}$ be a Poisson process with continuous intensity function $(\lambda(t))_{t \geq 0}$.

(a) Show that the intensities $\lambda_{n,n+k}(t)$, $n, k \geq 0$ and $t > 0$, of the Markov process N with transition probabilities $p_{n,n+k}(s,t)$ exist, i.e.,

$$\lambda_{n,n+k}(t) = \lim_{h \downarrow 0} \frac{p_{n,n+k}(t, t+h)}{h}, \quad n \geq 0, k \geq 1,$$

and that they are given by

$$\lambda_{n,n+k}(t) = \begin{cases} \lambda(t), & k = 1, \\ 0, & k \geq 2. \end{cases} \tag{2.1.25}$$

(b) What can you conclude from $p_{n,n+k}(t, t+h)$ for h small about the short term jump behavior of the Markov process N?

(c) Show by counterexample that (2.1.25) is in general not valid if one gives up the assumption of continuity of the intensity function $\lambda(t)$.

(2) Let $N = (N(t))_{t \geq 0}$ be a Poisson process with continuous intensity function $(\lambda(t))_{t \geq 0}$. By using the properties of N given in Definition 2.1.1, show that the following properties hold:

(a) The sample paths of N are non-decreasing.

(b) The process N does not have a jump at zero with probability 1.

(c) For every fixed t, the process N does not have a jump at t with probability 1. Does this mean that the sample paths do not have jumps?

(3) Let N be a homogeneous Poisson process on $[0, \infty)$. Show that for $0 < t_1 < t < t_2$,

$$\lim_{h \downarrow 0} P(N(t_1 - h, t - h] = 0, N(t - h, t] = 1, N(t, t_2] = 0 \mid N(t - h, t] > 0)$$

$$= e^{-\lambda(t - t_1)} e^{-\lambda(t_2 - t)}.$$

Give an intuitive interpretation of this property.

(4) Let $N_1, \ldots, N_n$ be independent Poisson processes on $[0, \infty)$ defined on the same probability space. Show that $N_1 + \cdots + N_n$ is a Poisson process and determine its mean value function.

This property extends the well-known property that the sum $M_1 + M_2$ of two independent Poisson random variables $M_1 \sim \text{Pois}(\lambda_1)$ and $M_2 \sim \text{Pois}(\lambda_2)$ is $\text{Pois}(\lambda_1 + \lambda_2)$. We also mention that a converse to this result holds. Indeed, suppose $M = M_1 + M_2$, $M \sim \text{Pois}(\lambda)$ for some $\lambda > 0$ and M_1, M_2 are independent non-negative random variables. Then both M_1 and M_2 are necessarily Poisson random variables. This phenomenon is referred to as *Raikov's theorem*; see Lukacs [54], Theorem 8.2.2. An analogous theorem can be shown for so-called *point processes* which are counting processes on $[0, \infty)$, including the Poisson process and the renewal process. Indeed, if the Poisson process N has representation $N \overset{d}{=} N_1 + N_2$ for independent point processes N_1, N_2, then N_1 and N_2 are necessarily Poisson processes.

(5) Consider the total claim amount process S in the Cramér-Lundberg model.

(a) Show that the total claim amount $S(s, t]$ in $(s, t]$ for $s < t$, i.e., $S(s, t] = S(t) - S(s)$, has the same distribution as the total claim amount in $[0, t - s]$, i.e., $S(t - s)$.

(b) Show that, for every $0 = t_0 < t_1 < \cdots < t_n$ and $n \geq 1$, the random variables $S(t_1), S(t_1, t_2], \ldots, S(t_{n-1}, t_n]$ are independent. Hint: Calculate the joint characteristic function of the latter random variables.

(6) For a homogeneous Poisson process N on $[0, \infty)$ show that for $0 < s < t$,

$$P(N(s) = k \mid N(t)) = \begin{cases} \dbinom{N(t)}{k} \left(\dfrac{s}{t}\right)^k \left(1 - \dfrac{s}{t}\right)^{N(t)-k} & \text{if } k \leq N(t), \\ 0 & \text{if } k > N(t). \end{cases}$$

Section 2.1.3

(7) Let $\widetilde{N}$ be a standard homogeneous Poisson process on $[0, \infty)$ and N a Poisson process on $[0, \infty)$ with mean value function μ.

(a) Show that $N_1(t) = (\widetilde{N}(\mu(t)))_{t \geq 0}$ is a Poisson process on $[0, \infty)$ with mean value function μ.

(b) Assume that the inverse μ^{-1} of μ exists, is continuous and $\lim_{t \to \infty} \mu(t) = \infty$. Show that $\widetilde{N}_1(t) = N(\mu^{-1}(t))$ defines a standard homogeneous Poisson process on $[0, \infty)$.

(c) Assume that the Poisson process N has an intensity function λ. Which condition on λ ensures that $\mu^{-1}(t)$ exists for $t \geq 0$?

(d) Let $f : [0, \infty) \to [0, \infty)$ be a non-decreasing continuous function with $f(0) = 0$. Show that

$$N_f(t) = N(f(t)), \quad t \geq 0,$$

is again a Poisson process on $[0, \infty)$. Determine its mean value function.

Sections 2.1.4-2.1.5

(8) The homogeneous Poisson process $\widetilde{N}$ with intensity $\lambda > 0$ can be written as a renewal process

$$\widetilde{N}(t) = \#\{i \geq 1 : \widetilde{T}_i \leq t\}, \quad t \geq 0,$$

where $\widetilde{T}_n = \widetilde{W}_1 + \cdots + \widetilde{W}_n$ and $(\widetilde{W}_n)$ is an iid $\text{Exp}(\lambda)$ sequence. Let N be a Poisson process with mean value function μ which has an a.e. positive continuous intensity function λ. Let $0 \leq T_1 \leq T_2 \leq \cdots$ be the arrival times of the process N.

(a) Show that the random variables $\int_{T_n}^{T_{n+1}} \lambda(s) \, ds$ are iid exponentially distributed.

(b) Show that, with probability 1, no multiple claims can occur, i.e., at an arrival time T_i of a claim, $N(T_i) - N(T_i-) = 1$ a.s. and $P(N(T_i) - N(T_i-) > 1$ for some $i) = 0$.

(9) Consider a homogeneous Poisson process N with intensity $\lambda > 0$ and arrival times T_i.

(a) Assume the renewal representation $N(t) = \#\{i \geq 1 : T_i \leq t\}$, $t \geq 0$, for N, i.e., $T_0 = 0$, $W_i = T_i - T_{i-1}$ are iid $\text{Exp}(\lambda)$ inter-arrival times. Calculate for $0 \leq t_1 < t_2$,

$$P(T_1 \leq t_1) \quad \text{and} \quad P(T_1 \leq t_1, T_2 \leq t_2). \qquad (2.1.26)$$

(b) Assume the properties of Definition 2.1.1 for N. Calculate for $0 \leq t_1 < t_2$,

$$P(N(t_1) \geq 1) \quad \text{and} \quad P(N(t_1) \geq 1, N(t_2) \geq 2). \qquad (2.1.27)$$

(c) Give reasons why you get the same probabilities in (2.1.26) and (2.1.27).

(10) Consider a homogeneous Poisson process on $[0, \infty)$ with arrival time sequence (T_i) and set $T_0 = 0$. The inter-arrival times are defined as $W_i = T_i - T_{i-1}$, $i \geq 1$.

(a) Show that T_1 has the *forgetfulness property*, i.e., $P(T_1 > t + s \mid T_1 > t) = P(T_1 > s)$, $t, s \geq 0$.

(b) Another version of the forgetfulness property is as follows. Let $Y \geq 0$ be independent of T_1 and Z be a random variable whose distribution is given by

$$P(Z > z) = P(T_1 > Y + z \mid T_1 > Y), \quad z \geq 0.$$

Then Z and T_1 have the same distribution. Verify this.

(c) Show that the events $\{W_1 < W_2\}$ and $\{\min(W_1, W_2) > x\}$ are independent.

(d) Determine the distribution of $m_n = \min(T_1, T_2 - T_1, \ldots, T_n - T_{n-1})$.

(11) Suppose you want to simulate sample paths of a Poisson process.

(a) How can you exploit the renewal representation to simulate paths of a homogeneous Poisson process?

(b) How can you use the renewal representation of a homogeneous Poisson N to simulate paths of an inhomogeneous Poisson process?

Sections 2.1.6

(12) Let $U_1, \ldots, U_n$ be an iid $U(0, 1)$ sample with the corresponding order statistics $U_{(1)} < \cdots < U_{(n)}$ a.s. Let $(\widetilde{W}_i)$ be an iid sequence of $\text{Exp}(\lambda)$ distributed random variables and $\widetilde{T}_n = \widetilde{W}_1 + \cdots + \widetilde{W}_n$ the corresponding arrival times of a homogeneous Poisson process with intensity λ.

(a) Show that the following identity in distribution holds for every fixed $n \geq 1$:

$$\left(U_{(1)}, \ldots, U_{(n)}\right) \stackrel{d}{=} \left(\frac{\widetilde{T}_1}{\widetilde{T}_{n+1}}, \ldots, \frac{\widetilde{T}_n}{\widetilde{T}_{n+1}}\right). \qquad (2.1.28)$$

Hint: Calculate the densities of the vectors on both sides of (2.1.28). The density of the vector

$$[(\widetilde{T}_1, \ldots, \widetilde{T}_n)/\widetilde{T}_{n+1}, \widetilde{T}_{n+1}]$$

can be obtained from the known density of the vector $(\widetilde{T}_1, \ldots, \widetilde{T}_{n+1})$.

(b) Why is the distribution of the right-hand vector in (2.1.28) independent of λ?

(c) Let T_i be the arrivals of a Poisson process on $[0, \infty)$ with a.e. positive intensity function λ and mean value function μ. Show that the following identity in distribution holds for every fixed $n \geq 1$:

$$\left(U_{(1)}, \dots, U_{(n)}\right) \overset{d}{=} \left(\frac{\mu(T_1)}{\mu(T_{n+1})}, \dots, \frac{\mu(T_n)}{\mu(T_{n+1})}\right).$$

(13) Let $W_1, \dots, W_n$ be an iid $\mathrm{Exp}(\lambda)$ sample for some $\lambda > 0$. Show that the ordered sample $W_{(1)} < \dots < W_{(n)}$ has representation in distribution:

$$\left(W_{(1)}, \dots, W_{(n)}\right)$$

$$\overset{d}{=} \left(\frac{W_n}{n}, \frac{W_n}{n} + \frac{W_{n-1}}{n-1}, \dots, \frac{W_n}{n} + \frac{W_{n-1}}{n-1} + \dots + \frac{W_2}{2},\right.$$

$$\left.\frac{W_n}{n} + \frac{W_{n-1}}{n-1} + \dots + \frac{W_1}{1}\right).$$

Hint: Use a density transformation starting with the joint density of $W_1, \dots, W_n$ to determine the density of the right-hand expression.

(14) Consider the stochastically discounted total claim amount

$$S(t) = \sum_{i=1}^{N(t)} e^{-rT_i} X_i,$$

where $r > 0$ is an interest rate, $0 < T_1 < T_2 < \cdots$ are the claim arrival times, defining the homogeneous Poisson process $N(t) = \#\{i \geq 1 : T_i \leq t\}$, $t \geq 0$, with intensity $\lambda > 0$, and (X_i) is an iid sequence of positive claim sizes, independent of (T_i).

(a) Calculate the mean and the variance of $S(t)$ by using the order statistics property of the Poisson process N. Specify the mean and the variance in the case when $r = 0$ (Cramér-Lundberg model).

(b) Show that $S(t)$ has the same distribution as

$$e^{-rt} \sum_{i=1}^{N(t)} e^{rT_i} X_i.$$

(15) Suppose you want to simulate sample paths of a Poisson process on $[0, T]$ for $T > 0$ and a given continuous intensity function λ, by using the order statistics property.

(a) How should you proceed if you are interested in one path with exactly n jumps in $[0, T]$?

(b) How would you simulate several paths of a homogeneous Poisson process with (possibly) different jump numbers in $[0, T]$?

(c) How could you use the simulated paths of a homogeneous Poisson process to obtain the paths of an inhomogeneous one with given intensity function?

(16) Let (T_i) be the arrival sequence of a standard homogeneous Poisson process N and $\alpha \in (0, 1)$.

(a) Show that the infinite series

$$X_\alpha = \sum_{i=1}^{\infty} T_i^{-1/\alpha} \qquad (2.1.29)$$

converges a.s. Hint: Use the strong law of large numbers for (T_n).

(b) Show that

$$X_{N(t)} = \sum_{i=1}^{N(t)} T_i^{-1/\alpha} \stackrel{a.s.}{\to} X_\alpha \quad \text{as } t \to \infty.$$

Hint: Use Lemma 2.2.6.

(c) It follows from standard limit theory for sums of iid random variables (see Feller [32], Theorem 1 in Chapter XVII.5) that for iid $U(0,1)$ random variables U_i,

$$n^{-1/\alpha} \sum_{i=1}^{n} U_i^{-1/\alpha} \stackrel{d}{\to} Z_\alpha, \qquad (2.1.30)$$

where Z_α is a positive random variable with an α-*stable distribution* determined by its Laplace-Stieltjes transform $E \exp\{-s\, Z_\alpha\} = \exp\{-c\, s^\alpha\}$ for some $c > 0$, all $s \geq 0$. See p. 182 for some information about Laplace-Stieltjes transforms.
Show that $X_\alpha \stackrel{d}{=} c'\, Z_\alpha$ for some positive constant $c' > 0$.
Hints: (i) Apply the order statistics property of the homogeneous Poisson process to $X_{N(t)}$ to conclude that

$$X_{N(t)} \stackrel{d}{=} t^{-1/\alpha} \sum_{i=1}^{N(t)} U_i^{-1/\alpha},$$

where (U_i) is an iid $U(0,1)$ sequence, independent of $N(t)$.
(ii) Prove that

$$(N(t))^{-1/\alpha} \sum_{i=1}^{N(t)} U_i^{-1/\alpha} \stackrel{d}{\to} Z_\alpha \quad \text{as } t \to \infty.$$

Hint: Condition on $N(t)$ and exploit (2.1.30).
(iii) Use the strong law of large numbers $N(t)/t \stackrel{a.s.}{\to} 1$ as $t \to \infty$ (Theorem 2.2.4) and the continuous mapping theorem to conclude the proof.

(d) Show that $EX_\alpha = \infty$.

(e) Let $Z_1, \ldots, Z_n$ be iid copies of the α-stable random variable Z_α with Laplace-Stieltjes transform $Ee^{-s\, Z_\alpha} = e^{-c\, s^\alpha}$, $s \geq 0$, for some $\alpha \in (0,1)$ and $c > 0$. Show that for every $n \geq 1$ the relation

$$Z_1 + \cdots + Z_n \stackrel{d}{=} n^{1/\alpha}\, Z_\alpha$$

holds. It is due to this "stability condition" that the distribution gained its name.
Hint: Use the properties of Laplace-Stieltjes transforms (see p. 182) to show this property.

(f) Consider Z_α from (e) for some $\alpha \in (0,1)$.

(i) Show the relation

$$Ee^{itAZ_\alpha^{1/2}} = e^{-c|t|^{2\alpha}}, \quad t \in \mathbb{R}, \tag{2.1.31}$$

where $A \sim N(0,2)$ is independent of Z_α. A random Y with characteristic function given by the right-hand side of (2.1.31) and its distribution are said to be *symmetric 2α-stable*.

(ii) Let $Y_1, \ldots, Y_n$ be iid copies of Y from (i). Show the stability relation

$$Y_1 + \cdots + Y_n \overset{d}{=} n^{1/(2\alpha)}\, Y\,.$$

(iii) Conclude that Y must have infinite variance. Hint: Suppose that Y has finite variance and try to apply the central limit theorem.

The interested reader who wants to learn more about the exciting class of stable distributions and stable processes is referred to Samorodnitsky and Taqqu [70].

Section 2.1.8

(17) Let $(N(t))_{t\geq 0}$ be a standard homogeneous Poisson process with claim arrival times T_i.

(a) Show that the sequences of arrival times $(\sqrt{T_i})$ and (T_i^2) define two Poisson processes N_1 and N_2, respectively, on $[0,\infty)$. Determine their mean measures by calculating $EN_i(s,t]$ for any $s < t$, $i = 1, 2$.

(b) Let N_3 and N_4 be Poisson processes on $[0,\infty)$ with mean value functions $\mu_3(t) = \sqrt{t}$ and $\mu_4(t) = t^2$ and arrival time sequences $(T_i^{(3)})$ and $(T_i^{(4)})$, respectively. Show that the processes $(N_3(t^2))_{t\geq 0}$ and $(N_4(\sqrt{t}))_{t\geq 0}$ are Poisson on $[0,\infty)$ and have the same distribution.

(c) Show that the process

$$N_5(t) = \#\{i \geq 1 : e^{T_i} \leq t+1\}, \quad t \geq 0\,,$$

is a Poisson process and determine its mean value function.

(d) Let N_6 be a Poisson process on $[0,\infty)$ with mean value function $\mu_6(t) = \log(1 + t)$. Show that N_6 has the property that, for $1 \leq s < t$ and $a \geq 1$, the distribution of $N_6(at-1) - N_6(as-1)$ does not depend on a.

(18) Let (T_i) be the arrival times of a homogeneous Poisson process N on $[0,\infty)$ with intensity $\lambda > 0$, independent of the iid claim size sequence (X_i) with $X_i > 0$ and distribution function F.

(a) Show that for $s < t$ and $a < b$ the counting random variable

$$M((s,t] \times (a,b]) = \#\{i \geq 1 : T_i \in (s,t]\,, X_i \in (a,b]\}$$

is $\mathrm{Pois}(\lambda(t-s)F(a,b])$ distributed.

(b) Let $\Delta_i = (s_i, t_i] \times (a_i, b_i]$ for $s_i < t_i$ and $a_i < b_i$, $i = 1, 2$, be disjoint. Show that $M(\Delta_1)$ and $M(\Delta_2)$ are independent.

(19) Consider the two-dimensional PRM $N_{\tilde{\psi}}$ from Figure 2.1.30 with $\alpha > 0$.

(a) Calculate the mean measure of the set $A(r,S) = \{\mathbf{x} : |\mathbf{x}| > r\,, \mathbf{x}/|\mathbf{x}| \in S\}$, where $r > 0$ and S is any Borel subset of the unit circle.

(b) Show that $EN_{\tilde{\psi}}(A(rt,S)) = t^{-\alpha}\, EN_{\tilde{\psi}}(A(r,S))$ for any $t > 0$.

(c) Let $\mathbf{Y} = R(\cos(2\pi X), \sin(2\pi X))$, where $P(R > x) = x^{-\alpha}$, $x \geq 1$, X is uniformly distributed on $(0,1)$ and independent of R. Show that for $r \geq 1$,

$$EN_{\tilde{\psi}}(A(r,S)) = P(\mathbf{Y} \in A(r,S))\,.$$

(20) Let $(E, \mathcal{E}, \mu)$ be a measure space such that $0 < \mu(E) < \infty$ and τ be $\mathrm{Pois}(\mu(E))$ distributed. Assume that τ is independent of the iid sequence (X_i) with distribution given by

$$F_{X_1}(A) = P(X_1 \in A) = \mu(A)/\mu(E), \quad A \in \mathcal{E}.$$

(a) Show that the counting process

$$N(A) = \sum_{i=1}^{\tau} I_A(X_i), \quad A \in \mathcal{E},$$

is $\mathrm{PRM}(\mu)$ on E. Hint: Calculate the joint characteristic function of the random variables $N(A_1), \ldots, N(A_m)$ for any disjoint $A_1, \ldots, A_m \in \mathcal{E}$.

(b) Specify the construction of (a) in the case that $E = [0, 1]$ equipped with the Borel σ-field, when μ has an a.e. positive density λ. What is the relation with the order statistics property of the Poisson process N?

(c) Specify the construction of (a) in the case that $E = [0, 1]^d$ equipped with the Borel σ-field for some integer $d \geq 1$ when $\mu = \lambda\,\mathrm{Leb}$ for some constant $\lambda > 0$. Propose how one could define an "order statistics property" for this (homogeneous) Poisson process with points in E.

(21) Let τ be a $\mathrm{Pois}(1)$ random variable, independent of the iid sequence (X_i) with common distribution function F and a positive density on $(0, \infty)$.

(a) Show that

$$N(t) = \sum_{i=1}^{\tau} I_{(0,t]}(X_i), \quad t \geq 0,$$

defines a Poisson process on $[0, \infty)$ in the sense of Definition 2.1.1.

(b) Determine the mean value function of N.

(c) Find a function $f : [0, \infty) \to [0, \infty)$ such that the time changed process $(N(f(t)))_{t \geq 0}$ becomes a standard homogeneous Poisson process.

(22) For an iid sequence (X_i) with common continuous distribution function F define the sequence of partial maxima $M_n = \max(X_1, \ldots, X_n)$, $n \geq 1$. Define $L(1) = 1$ and, for $n \geq 1$,

$$L(n + 1) = \inf\{k > L(n) : X_k > X_{L(n)}\}.$$

The sequence $(X_{L(n)})$ is called the *record value sequence* and $(L(n))$ is the sequence of the *record times*.

It is well-known that for an iid standard exponential sequence (W_i) with record time sequence $(\widetilde{L}(n))$, $(W_{\widetilde{L}(n)})$ constitute the arrivals of a standard homogeneous Poisson process on $[0, \infty)$; see Resnick [64], Proposition 4.1.

(a) Let $R(x) = -\log \overline{F}(x)$, where $\overline{F} = 1 - F$ and $x \in (x_l, x_r)$, $x_l = \inf\{x : F(x) > 0\}$ and $x_r = \sup\{x : F(x) < 1\}$. Show that $(X_{L(n)}) \overset{d}{=} (R^{\leftarrow}(W_{\widetilde{L}(n)}))$, where $R^{\leftarrow}(t) = \inf\{x \in (x_l, x_r) : R(x) \geq t\}$ is the *generalized inverse of R*. See Resnick [64], Proposition 4.1.

(b) Conclude from (a) that $(X_{L(n)})$ is the arrival sequence of a Poisson process on (x_l, x_r) with mean measure of $(a, b] \subset (x_l, x_r)$ given by $R(a, b]$.

2.2 The Renewal Process

2.2.1 Basic Properties

In Section 2.1.4 we learned that the homogeneous Poisson process is a particular renewal process. In this section we want to study this model. We start with a formal definition.

Definition 2.2.1 (Renewal process)
Let (W_i) be an iid sequence of a.s. positive random variables. Then the random walk

$$T_0 = 0, \quad T_n = W_1 + \cdots + W_n, \quad n \geq 1,$$

is said to be a renewal sequence *and the counting process*

$$N(t) = \#\{i \geq 1 : T_i \leq t\} \quad t \geq 0,$$

is the corresponding renewal (counting) process.

We also refer to (T_n) and (W_n) as the sequences of the arrival and inter-arrival times of the renewal process N, respectively.

Example 2.2.2 (Homogeneous Poisson process)
It follows from Theorem 2.1.6 that a homogeneous Poisson process with intensity λ is a renewal process with iid exponential $\mathrm{Exp}(\lambda)$ inter-arrival times W_i. □

A main motivation for introducing the renewal process is that the (homogeneous) Poisson process does not always describe claim arrivals in an adequate way. There can be large gaps between arrivals of claims. For example, it is unlikely that windstorm claims arrive according to a homogeneous Poisson process. They happen now and then, sometimes with years in between. In this case it is more natural to assume that the inter-arrival times have a distribution which allows for modeling these large time intervals. The log-normal or the Pareto distributions would do this job since their tails are much heavier than those of the exponential distribution; see Section 3.2. We have also seen in Section 2.1.7 that the Poisson process is not always a realistic model for real-life claim arrivals, in particular if one considers long periods of time.

On the other hand, if we give up the hypothesis of a Poisson process we lose most of the nice properties of this process which are closely related to the exponential distribution of the W_i's. For example, it is in general unknown which distribution $N(t)$ has and what the exact values of $EN(t)$ or $\mathrm{var}(N(t))$ are. We will, however, see that the renewal processes and the homogeneous Poisson process have various *asymptotic* properties in common.

The first result of this kind is a strong law of large numbers for the renewal counting process.

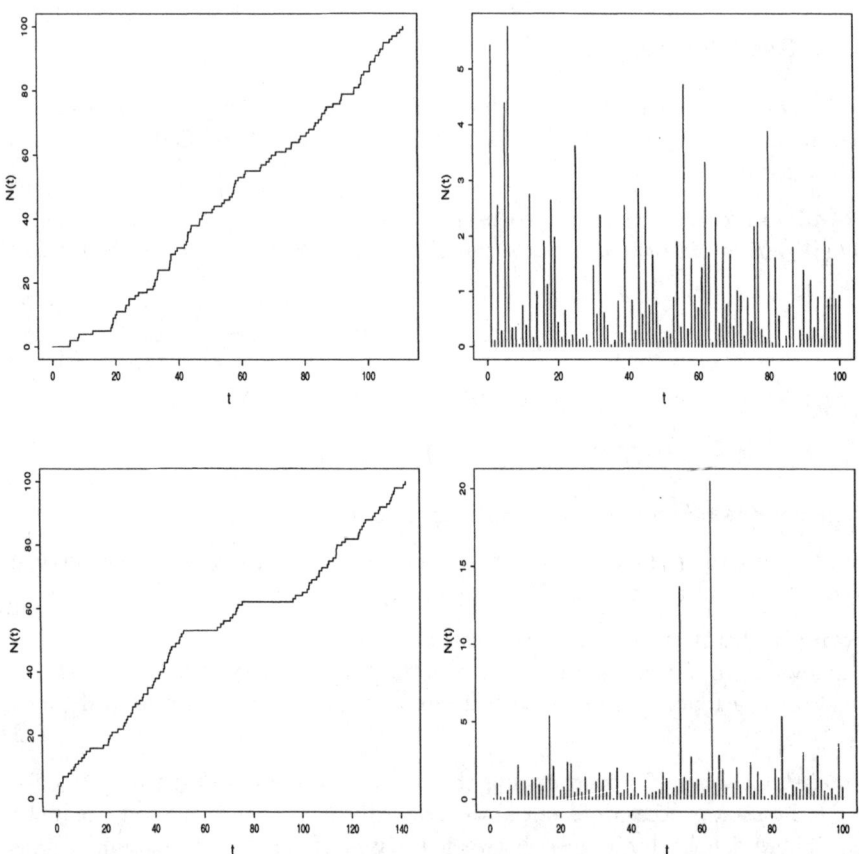

Figure 2.2.3 *One path of a renewal process* (left graphs) *and the corresponding renewal sequence* (right graphs). *Top: Standard homogeneous Poisson process with iid standard exponential inter-arrival times. Bottom: The renewal process has iid Pareto distributed inter-arrival times with* $P(W_i > x) = x^{-4}$, $x \geq 1$. *Both renewal paths have 100 jumps. Notice the extreme lengths of some inter-arrival times in the bottom graph; they are atypical for a homogeneous Poisson process.*

Theorem 2.2.4 (Strong law of large numbers for the renewal process)
If the expectation $EW_1 = \lambda^{-1}$ *of the inter-arrival times* W_i *is finite, N satisfies the strong law of large numbers:*

$$\lim_{t \to \infty} \frac{N(t)}{t} = \lambda \quad \text{a.s.}$$

Proof. We need a simple auxiliary result.

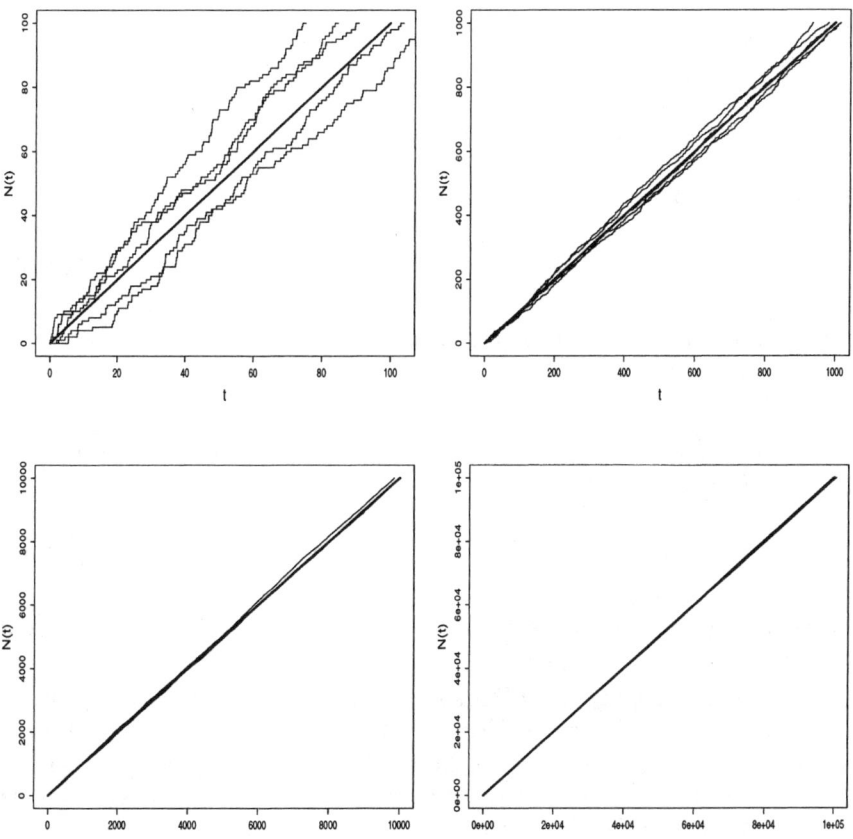

Figure 2.2.5 *Five paths of a renewal process with* $\lambda = 1$ *and* $n = 10^i$ *jumps,* $i = 2, 3, 4, 5$. *The mean value function* $EN(t) = t$ *is also indicated (solid straight line). The approximation of* $N(t)$ *by* $EN(t)$ *for increasing* t *is nicely illustrated; on a large time scale* $N(t)$ *and* $EN(t)$ *can hardly be distinguished.*

Lemma 2.2.6 *Let* (Z_n) *be a sequence of random variables such that* $Z_n \overset{\text{a.s.}}{\to} Z$ *as* $n \to \infty$ *for some random variable* Z, *and let* $(M(t))_{t \geq 0}$ *be a stochastic process of integer-valued random variables such that* $M(t) \overset{\text{a.s.}}{\to} \infty$ *as* $t \to \infty$. *If* M *and* (Z_n) *are defined on the same probability space* Ω, *then*

$$Z_{M(t)} \to Z \quad \text{a.s.} \quad as\ t \to \infty.$$

Proof. Write

$$\Omega_1 = \{\omega \in \Omega : M(t, \omega) \to \infty\} \quad \text{and} \quad \Omega_2 = \{\omega \in \Omega : Z_n(\omega) \to Z(\omega)\}.$$

By assumption, $P(\Omega_1) = P(\Omega_2) = 1$, hence $P(\Omega_1 \cap \Omega_2) = 1$ and therefore

$$P(\{\omega : Z_{M(t,\omega)}(\omega) \to Z\}) \geq P(\Omega_1 \cap \Omega_2) = 1.$$

This proves the lemma.   □

Recall the following basic relation of a renewal process:

$$\{N(t) = n\} = \{T_n \leq t < T_{n+1}\}, \quad n \in \mathbb{N}_0.$$

Then it is immediate that the following sandwich inequalities hold:

$$\frac{T_{N(t)}}{N(t)} \leq \frac{t}{N(t)} \leq \frac{T_{N(t)+1}}{N(t)+1} \frac{N(t)+1}{N(t)} \tag{2.2.32}$$

By the strong law of large numbers for the iid sequence (W_n) we have

$$n^{-1} T_n \overset{\text{a.s.}}{\to} \lambda^{-1}.$$

In particular, $N(t) \to \infty$ a.s. as $t \to \infty$. Now apply Lemma 2.2.6 with $Z_n = T_n/n$ and $M = N$ to obtain

$$\frac{T_{N(t)}}{N(t)} \overset{\text{a.s.}}{\to} \lambda^{-1}. \tag{2.2.33}$$

The statement of the theorem follows by a combination of (2.2.32) and (2.2.33).   □

In the case of a homogeneous Poisson process we know the exact value of the expected renewal process: $EN(t) = \lambda t$. In the case of a general renewal process N the strong law of large numbers $N(t)/t \overset{\text{a.s.}}{\to} \lambda = (EW_1)^{-1}$ suggests that the expectation $EN(t)$ of the renewal process is approximately of the order λt. A lower bound for $EN(t)/t$ is easily achieved. By an application of Fatou's lemma (see for example Williams [78]) and the strong law of large numbers for $N(t)$,

$$\lambda = E \liminf_{t \to \infty} \frac{N(t)}{t} \leq \liminf_{t \to \infty} \frac{EN(t)}{t}. \tag{2.2.34}$$

This lower bound can be complemented by the corresponding upper one which leads to the following standard result.

Theorem 2.2.7 (Elementary renewal theorem)
If the expectation $EW_1 = \lambda^{-1}$ of the inter-arrival times is finite, the following relation holds:

$$\lim_{t \to \infty} \frac{EN(t)}{t} = \lambda.$$

Proof. By virtue of (2.2.34) it remains to prove that

$$\limsup_{t \to \infty} \frac{EN(t)}{t} \leq \lambda. \tag{2.2.35}$$

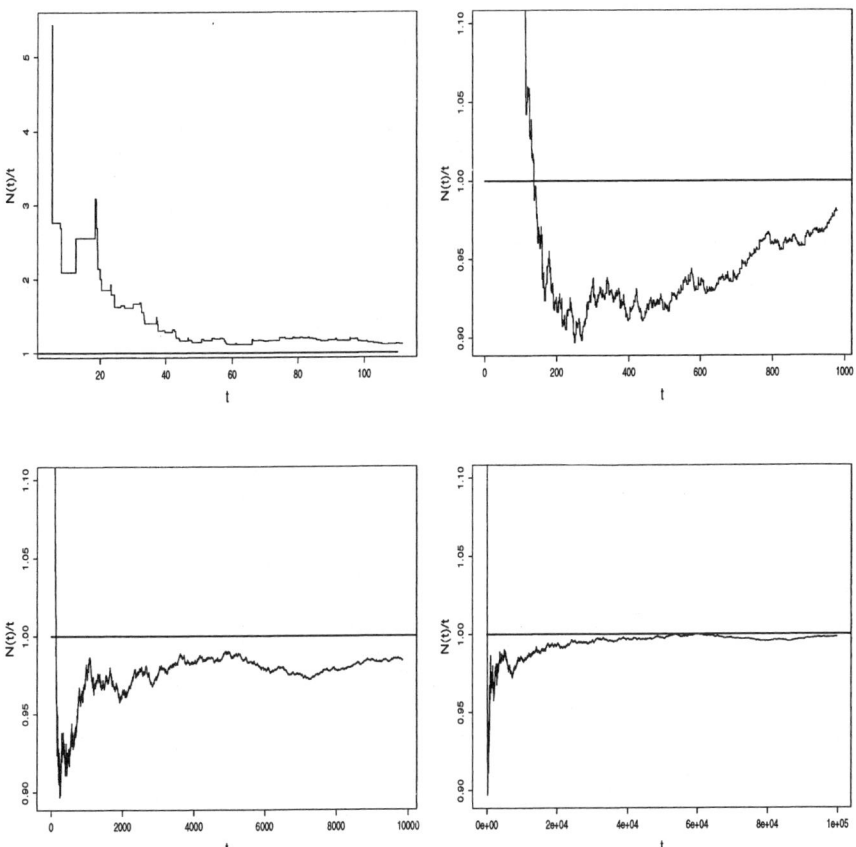

Figure 2.2.8 *The ratio $N(t)/t$ for a renewal process with $n = 10^i$ jumps, $i = 2, 3, 4, 5$, and $\lambda = 1$. The strong law of large numbers forces $N(t)/t$ towards 1 for large t.*

We use a truncation argument which we borrow from Resnick [65], p. 191. Write for any $b > 0$,

$$W_i^{(b)} = \min(W_i, b)\,, \quad T_i^{(b)} = W_1^{(b)} + \cdots + W_i^{(b)}\,, \quad i \geq 1\,.$$

Obviously, $(T_n^{(b)})$ is a renewal sequence and $T_n \geq T_n^{(b)}$ which implies $N_b(t) \geq N(t)$ for the corresponding renewal process

$$N_b(t) = \#\{i \geq 1 : T_i^{(b)} \leq t\}\,, \quad t \geq 0\,.$$

Hence

$$\limsup_{t \to \infty} \frac{EN(t)}{t} \leq \limsup_{t \to \infty} \frac{EN_b(t)}{t}\,. \tag{2.2.36}$$

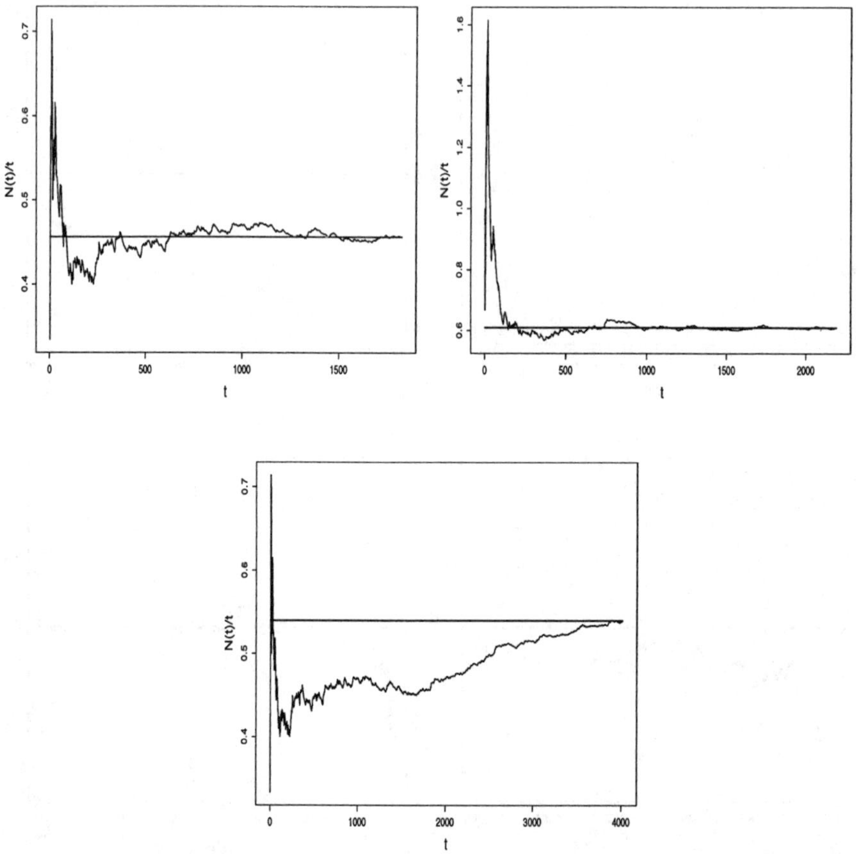

Figure 2.2.9 *Visualization of the validity of the strong law of large numbers for the arrivals of the Danish fire insurance data* 1980 − 1990; *see Section 2.1.7 for a description of the data.* Top left: *The ratio $N(t)/t$ for* 1980 − 1984, *where $N(t)$ is the claim number at day t in this period. The values cluster around the value* 0.46 *which is indicated by the constant line.* Top right: *The ratio $N(t)/t$ for* 1985 − 1990, *where $N(t)$ is the claim number at day t in this period. The values cluster around the value* 0.61 *which is indicated by the constant line.* Bottom: *The ratio $N(t)/t$ for the whole period* 1980 − 1990, *where $N(t)$ is the claim number at day t in this period. The graph gives evidence about the fact that the strong law of large numbers does not apply to N for the whole period. This is caused by an increase of the annual intensity in* 1985 − 1990 *which can be observed in Figure 2.1.21. This fact makes the assumption of iid inter-arrival times over the whole period of* 11 *years questionable. We do, however, see in the top graphs that the strong law of large numbers works satisfactorily in the two distinct periods.*

We observe that, by definition of N_b,

$$T^{(b)}_{N_b(t)} = W^{(b)}_1 + \cdots + W^{(b)}_{N_b(t)} \le t.$$

The following result is due to the fact that $N_b(t) + 1$ is a so-called *stopping time*[19] *with respect to the natural filtration* generated by the sequence $(W^{(b)}_i)$. Then the relation

$$E(T^{(b)}_{N_b(t)+1}) = E(N_b(t) + 1)\, EW^{(b)}_1 \tag{2.2.37}$$

holds by virtue of *Wald's identity*. Combining (2.2.36)-(2.2.37), we conclude that

$$\limsup_{t\to\infty} \frac{EN(t)}{t} \le \limsup_{t\to\infty} \frac{E(T^{(b)}_{N_b(t)})}{t\, EW^{(b)}_1} \le \limsup_{t\to\infty} \frac{t}{t\, EW^{(b)}_1} = (EW^{(b)}_1)^{-1}.$$

Since by the monotone convergence theorem (see for example Williams [78]), letting $b \uparrow \infty$,

$$EW^{(b)}_1 = E(\min(b, W_1)) \uparrow EW_1 = \lambda^{-1},$$

the desired relation (2.2.35) follows. This concludes the proof. $\qquad\square$

For further reference we include a result about the asymptotic behavior of $\mathrm{var}(N(t))$. The proof can be found in Gut [40], Theorem 5.2.

Proposition 2.2.10 (The asymptotic behavior of the variance of the renewal process)
Assume $\mathrm{var}(W_1) < \infty$. *Then*

$$\lim_{t\to\infty} \frac{\mathrm{var}(N(t))}{t} = \frac{\mathrm{var}(W_1)}{(EW_1)^3}.$$

Finally, we mention that $N(t)$ satisfies the central limit theorem; see Embrechts et al. [29], Theorem 2.5.13, for a proof.

Theorem 2.2.11 (The central limit theorem for the renewal process)
Assume that $\mathrm{var}(W_1) < \infty$. *Then the central limit theorem*

$$(\mathrm{var}(W_1)\,(EW_1)^{-3}\,t)^{-1/2}\,(N(t) - \lambda t) \stackrel{d}{\to} Y \sim N(0,1). \tag{2.2.38}$$

holds as $t \to \infty$.

---

[19] Let $\mathcal{F}_n = \sigma(W^{(b)}_i, i \le n)$ be the σ-field generated by $W^{(b)}_1, \ldots, W^{(b)}_n$. Then $(\mathcal{F}_n)$ is the natural filtration generated by the sequence $(W^{(b)}_n)$. An integer-valued random variable τ is a stopping time with respect to $(\mathcal{F}_n)$ if $\{\tau = n\} \in \mathcal{F}_n$. If $E\tau < \infty$ Wald's identity yields $E\left(\sum_{i=1}^{\tau} W^{(b)}_i\right) = E\tau\, EW^{(b)}_1$. Notice that $\{N_b(t) = n\} = \{T^{(b)}_n \le t < T^{(b)}_{n+1}\}$. Hence $N_b(t)$ is not a stopping time. However, the same argument shows that $N_b(t) + 1$ is a stopping time with respect to $(\mathcal{F}_n)$. The interested reader is referred to Williams's textbook [78] which gives a concise introduction to discrete-time martingales, filtrations and stopping times.

By virtue of Proposition 2.2.10, the normalizing constants $\sqrt{\mathrm{var}(W_1)}(EW_1)^{-3t}$ in (2.2.38) can be replaced by the standard deviation $\sqrt{\mathrm{var}(N(t))}$.

2.2.2 An Informal Discussion of Renewal Theory

Renewal processes model occurrences of events happening at random instants of time, where the inter-arrival times are approximately iid. In the context of non-life insurance these instants were interpreted as the arrival times of claims. Renewal processes play a major role in applied probability. Complex stochastic systems can often be described by one or several renewal processes as building blocks. For example, the Internet can be understood as the superposition of a huge number of ON/OFF processes. Each of these processes corresponds to one "source" (computer) which communicates with other sources. ON refers to an active period of the source, OFF to a period of silence. The ON/OFF periods of each source constitute two sequences of iid positive random variables, both defining renewal processes.[20] A renewal process is also defined by the sequence of renewals (times of replacement) of a technical device or tool, say the light bulbs in a lamp or the fuel in a nuclear power station. From these elementary applications the process gained its name.

Because of their theoretical importance renewal processes are among the best studied processes in applied probability theory. The object of main interest in renewal theory is the *renewal function*[21]

$$m(t) = EN(t) + 1, \quad t \geq 0.$$

It describes the average behavior of the renewal counting process. In the insurance context, this is the expected number of claim arrivals in a portfolio. This number certainly plays an important role in the insurance business and its theoretical understanding is therefore essential. The iid assumption of the inter-arrival times is perhaps not the most realistic but is convenient for building up a theory.

The *elementary renewal theorem* (Theorem 2.2.7) is a simple but not very precise result about the average behavior of renewals: $m(t) = \lambda t\,(1 + o(1))$ as $t \to \infty$, provided $EW_1 = \lambda^{-1} < \infty$. Much more precise information is gained by *Blackwell's renewal theorem*. It says that for $h > 0$,

$$m(t, t + h] = EN(t, t + h] \to \lambda h, \quad t \to \infty.$$

---

[20] The approach to tele-traffic via superpositions of ON/OFF processes became popular in the 1990s; see Willinger et al. [79].

[21] The addition of one unit to the mean $EN(t)$ refers to the fact that $T_0 = 0$ is often considered as the first renewal time. This definition often leads to more elegant theoretical formulations. Alternatively, we have learned on p. 65 that the process $N(t) + 1$ has the desirable theoretical property of a stopping time, which $N(t)$ does not have.

(For Blackwell's renewal theorem and the further statements of this section we assume that the inter-arrival times W_i have a density.) Thus, for sufficiently large t, the expected number of renewals in the interval $(t, t + h]$ becomes independent of t and is proportional to the length of the interval. Since m is a non-decreasing function on $[0, \infty)$ it defines a measure m (we use the same symbol for convenience) on the Borel σ-field of $[0, \infty)$, the so-called *renewal measure*.

A special calculus has been developed for integrals with respect to the renewal measure. In this context, the crucial condition on the integrands is called *direct Riemann integrability*. Directly Riemann integrable functions on $[0, \infty)$ constitute quite a sophisticated class of integrands; it includes Riemann integrable functions on $[0, \infty)$ which have compact support (the function vanishes outside a certain finite interval) or which are non-increasing and non-negative. The *key renewal theorem* states that for a directly Riemann integrable function f,

$$\int_0^t f(t - s) \, dm(s) \to \lambda \int_0^\infty f(s) \, ds \,. \tag{2.2.39}$$

Under general conditions, it is equivalent to Blackwell's renewal theorem which, in a sense, is a special case of (2.2.39) for indicator functions $f(x) = I_{(0,h]}(x)$ with $h > 0$ and for $t > h$:

$$\int_0^t f(t - s) \, dm(s) = \int_{t-h}^t I_{(0,h]}(t - s) \, dm(s) = m(t - h, t]$$

$$\to \lambda \int_0^\infty f(s) \, ds = \lambda h \,.$$

An important part of renewal theory is devoted to the *renewal equation*. It is a convolution equation of the form

$$U(t) = u(t) + \int_0^t U(t - y) \, dF_{T_1}(y) \,, \tag{2.2.40}$$

where all functions are defined on $[0, \infty)$. The function U is *unknown*, u is a *known* function and F_{T_1} is the distribution function of the iid positive inter-arrival times $W_i = T_i - T_{i-1}$. The main goal is to find a solution U to (2.2.40). It is provided by the following general result which can be found in Resnick [65], p. 202.

Theorem 2.2.12 (W. Smith's key renewal theorem)

(1) *If u is bounded on every finite interval then*

$$U(t) = \int_0^t u(t - s) \, dm(s) \,, \quad t \geq 0 \,, \tag{2.2.41}$$

is the unique solution of the renewal equation (2.2.40) in the class of all functions on $(0, \infty)$ which are bounded on finite intervals. Here the right-hand integral has to be interpreted as $\int_{(-\infty,t]} u(t-s)\, dm(s)$ with the convention that $m(s) = u(s) = 0$ for $s < 0$.

(2) If, in addition, u is directly Riemann integrable, then

$$\lim_{t \to \infty} U(t) = \lambda \int_0^\infty u(s)\, ds\,.$$

Part (2) of the theorem is immediate from Blackwell's renewal theorem.

The renewal function itself satisfies the renewal equation with $u = I_{[0,\infty)}$. From this fact the general equation (2.2.40) gained its name.

Example 2.2.13 (The renewal function satisfies the renewal equation) Observe that for $t \geq 0$,

$$m(t) = EN(t) + 1 = 1 + E\left(\sum_{n=1}^\infty I_{[0,t]}(T_n)\right) = 1 + \sum_{n=1}^\infty P(T_n \leq t)$$

$$= I_{[0,\infty)}(t) + \sum_{n=1}^\infty \int_0^t P(y + (T_n - T_1) \leq t)\, dF_{T_1}(y)$$

$$= I_{[0,\infty)}(t) + \int_0^t \sum_{n=1}^\infty P(T_{n-1} \leq t - y)\, dF_{T_1}(y)$$

$$= I_{[0,\infty)}(t) + \int_0^t m(t - y)\, dF_{T_1}(y)\,.$$

This is a renewal equation with $U(t) = m(t)$ and $u(t) = I_{[0,\infty)}(t)$.   □

The usefulness of the renewal equation is illustrated in the following example.

Example 2.2.14 (Recurrence times of a renewal process)
In our presentation we closely follow Section 3.5 in Resnick [65]. Consider a renewal sequence (T_n) with $T_0 = 0$ and $W_n > 0$ a.s. Recall that

$$\{N(t) = n\} = \{T_n \leq t < T_{n+1}\}\,.$$

In particular, $T_{N(t)} \leq t < T_{N(t)+1}$. For $t \geq 0$, the quantities

$$F(t) = T_{N(t)+1} - t \quad \text{and} \quad B(t) = t - T_{N(t)}$$

are the *forward* and *backward recurrence times* of the renewal process, respectively. For obvious reasons, $F(t)$ is also called the *excess life* or *residual life*, i.e., it is the time until the next renewal, and $B(t)$ is called the *age process*. In an insurance context, $F(t)$ is the time until the next claim arrives, and $B(t)$ is the time which has evolved since the last claim arrived.

It is our aim to show that the function $P(B(t) \leq x)$ for fixed $0 \leq x < t$ satisfies a renewal equation. It suffices to consider the values $x < t$ since $B(t) \leq t$ a.s., hence $P(B(t) \leq x) = 1$ for $x \geq t$. We start with the identity

$$P(B(t) \leq x) = P(B(t) \leq x, T_1 \leq t) + P(B(t) \leq x, T_1 > t), \quad x > 0.$$

$$(2.2.42)$$

If $T_1 > t$, no jump has occurred by time t, hence $N(t) = 0$ and therefore $B(t) = t$. We conclude that

$$P(B(t) \leq x, T_1 > t) = (1 - F_{T_1}(t)) I_{[0,x]}(t). \qquad (2.2.43)$$

For $T_1 \leq t$, we want to show the following result:

$$P(B(t) \leq x, T_1 \leq t) = \int_0^t P(B(t-y) \leq x) \, dF_{T_1}(y). \qquad (2.2.44)$$

This means that, on the event $\{T_1 \leq t\}$, the process B "starts from scratch" at T_1. We make this precise by exploiting a "typical renewal argument". First observe that

$$P(B(t) \leq x, T_1 \leq t) = P(t - T_{N(t)} \leq x, N(t) \geq 1)$$

$$= \sum_{n=1}^{\infty} P(t - T_{N(t)} \leq x, N(t) = n)$$

$$= \sum_{n=1}^{\infty} P(t - T_n \leq x, T_n \leq t < T_{n+1}).$$

We study the summands individually by conditioning on $\{T_1 = y\}$ for $y \leq t$:

$$P(t - T_n \leq x, T_n \leq t < T_{n+1} \mid T_1 = y)$$

$$= P\left(t - \left[y + \sum_{i=2}^{n} W_i\right] \leq x, y + \sum_{i=2}^{n} W_i \leq t < y + \sum_{i=2}^{n+1} W_i\right)$$

$$= P\left(t - y - T_{n-1} \leq x, T_{n-1} \leq t - y \leq T_n\right)$$

$$= P\left(t - y - T_{N(t-y)} \leq x, N(t-y) = n - 1\right).$$

Hence we have

$$P(B(t) \leq x, T_1 \leq t)$$

$$= \sum_{n=0}^{\infty} \int_0^t P\left(t - y - T_{N(t-y)} \leq x, N(t-y) = n\right) dF_{T_1}(y)$$

$$= \int_0^t P(B(t-y) \leq x) \, dF_{T_1}(y),$$

which is the desired relation (2.2.44). Combining (2.2.42)-(2.2.44), we arrive at

$$P(B(t) \le x) = (1 - F_{T_1}(t)) I_{[0,x]}(t) + \int_0^t P(B(t-y) \le x) \, dF_{T_1}(y).$$

(2.2.45)

This is a renewal equation of the form (2.2.40) with $u(t) = (1-F_{T_1}(t)) I_{[0,x]}(t)$, and $U(t) = P(B(t) \le x)$ is the unknown function.

A similar renewal equation can be given for $P(F(t) > x)$:

$$P(F(t) > x) = \int_0^t P(F(t-y) > x) \, dF_{T_1}(y) + (1 - F_{T_1}(t+x)).$$

(2.2.46)

We mentioned before, see (2.2.41), that the unique solution to the renewal equation (2.2.45) is given by

$$U(t) = P(B(t) \le x) = \int_0^t (1 - F_{T_1}(t-y)) I_{[0,x]}(t-y) \, dm(y).$$

(2.2.47)

Now consider a homogeneous Poisson process with intensity λ. In this case, $m(t) = EN(t) + 1 = \lambda t + 1$, $1 - F_{T_1}(x) = \exp\{-\lambda x\}$. From (2.2.47) for $x < t$ and since $B(t) \le t$ a.s. we obtain

$$P(B(t) \le x) = P(t - T_{N(t)} \le x) = \begin{cases} 1 - e^{-\lambda x} & \text{if } x < t, \\ 1 & \text{if } x \ge t. \end{cases}$$

A similar argument yields for $F(t)$,

$$P(F(t) \le x) = P(T_{N(t)+1} - t \le x) = 1 - e^{-\lambda x}, \quad x > 0.$$

The latter result is counterintuitive in a sense since, on the one hand, the inter-arrival times W_i are $\text{Exp}(\lambda)$ distributed and, on the other hand, the time $T_{N(t)+1} - t$ until the next renewal has the same distribution. This reflects the *forgetfulness* property of the exponential distribution of the inter-arrival times. We refer to Example 2.1.7 for further discussions and a derivation of the distributions of $B(t)$ and $F(t)$ for the homogeneous Poisson process by elementary means. □

Comments

Renewal theory constitutes an important part of applied probability theory. Resnick [65] gives an entertaining introduction with various applications, among others, to problems of insurance mathematics. The advanced text on stochastic processes in insurance mathematics by Rolski et al. [67] makes extensive use of renewal techniques. Gut's book [40] is a collection of various useful limit results related to renewal theory and stopped random walks.

The notion of direct Riemann integrability has been discussed in various books; see Alsmeyer [1], p. 69, Asmussen [5], Feller [32], pp. 361-362, or Resnick [65], Section 3.10.1.

Smith's key renewal theorem will also be key to the asymptotic results on the ruin probability in the Cramér-Lundberg model in Section 4.2.2.

Exercises

(1) Let (T_i) be a renewal sequence with $T_0 = 0$, $T_n = W_1 + \cdots + W_n$, where (W_i) is an iid sequence of non-negative random variables.
(a) Which assumption is needed to ensure that the renewal process $N(t) = \#\{i \geq 1 : T_i \leq t\}$ has no jump sizes greater than 1 with positive probability?
(b) Can it happen that (T_i) has a limit point with positive probability? This would mean that $N(t) = \infty$ at some finite time t.
(2) Let N be a homogeneous Poisson process on $[0, \infty)$ with intensity $\lambda > 0$.
(a) Show that $N(t)$ satisfies the central limit theorem as $t \to \infty$ i.e.,

$$\widehat{N}(t) = \frac{N(t) - \lambda t}{\sqrt{\lambda t}} \xrightarrow{d} Y \sim N(0,1)\,,$$

(i) by using characteristic functions,
(ii) by employing the known central limit theorem for the sequence $((N(n) - \lambda n)/\sqrt{\lambda n})_{n=1,2,\ldots}$, and then by proving that $\max_{t \in (n,n+1]}(N(t) - N(n))/\sqrt{n} \xrightarrow{P} 0$.
(b) Show that N satisfies the multivariate central limit theorem for any $0 < s_1 < \cdots < s_n$ as $t \to \infty$:

$$(\sqrt{\lambda t})^{-1}\left(N(s_1 t) - s_1 \lambda t \ldots, N(s_n t) - s_n \lambda t\right) \xrightarrow{d} \mathbf{Y} \sim N(\mathbf{0}, \mathbf{\Sigma})\,,$$

where the right-hand distribution is multivariate normal with mean vector zero and covariance matrix $\mathbf{\Sigma}$ whose entries satisfy $\sigma_{i,j} = \min(s_i, s_j)$, $i, j = 1, \ldots, n$.
(3) Let $F(t) = T_{N(t)+1} - t$ be the forward recurrence time from Example 2.2.14.
(a) Show that the probability $P(F(t) > x)$, considered as a function of t, for $x > 0$ fixed satisfies the renewal equation (2.2.46).
(b) Solve (2.2.46) in the case of iid $\text{Exp}(\lambda)$ inter-arrival times.

2.3 The Mixed Poisson Process

In Section 2.1.3 we learned that an inhomogeneous Poisson process N with mean value function μ can be derived from a standard homogeneous Poisson process $\widetilde{N}$ by a deterministic time change. Indeed, the process

$$\widetilde{N}(\mu(t))\,, \quad t \geq 0\,,$$

has the same finite-dimensional distributions as N and is càdlàg, hence it is a possible representation of the process N. In what follows, we will use a similar construction by *randomizing the mean value function*.

Definition 2.3.1 (Mixed Poisson process)
Let $\widetilde{N}$ be a standard homogeneous Poisson process and μ be the mean value function of a Poisson process on $[0, \infty)$. Let $\theta > 0$ a.s. be a (non-degenerate) random variable independent of $\widetilde{N}$. Then the process

$$N(t) = \widetilde{N}(\theta \, \mu(t)), \quad t \geq 0,$$

is said to be a mixed Poisson process with mixing variable θ.

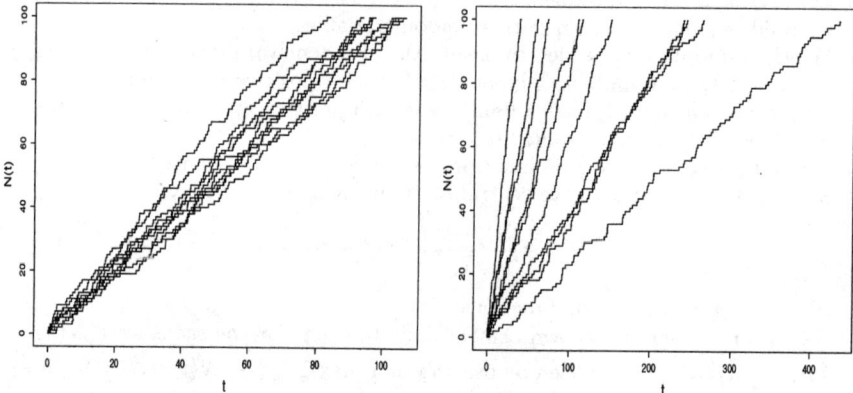

Figure 2.3.2 *Left: Ten sample paths of a standard homogeneous Poisson process. Right: Ten sample paths of a mixed homogeneous Poisson process with $\mu(t) = t$. The mixing variable θ is standard exponentially distributed. The processes in the left and right graphs have the same mean value function $EN(t) = t$.*

Example 2.3.3 (The negative binomial process as mixed Poisson process)
One of the important representatives of mixed Poisson processes is obtained by choosing $\mu(t) = t$ and θ gamma distributed. First recall that a $\Gamma(\gamma, \beta)$ distributed random variable θ has density

$$f_\theta(x) = \frac{\beta^\gamma}{\Gamma(\gamma)} \, x^{\gamma-1} \, e^{-\beta x}, \quad x > 0. \tag{2.3.48}$$

Also recall that an integer-valued random variable Z is said to be negative binomially distributed with parameter (p, v) if it has individual probabilities

$$P(Z = k) = \binom{v + k - 1}{k} p^v \, (1 - p)^k, \quad k \in \mathbb{N}_0, \quad p \in (0, 1), \quad v > 0.$$

Verify that $N(t)$ is negative binomial with parameter $(p, v) = (\beta/(t+\beta), \gamma)$. $\square$

In an insurance context, a mixed Poisson process is introduced as a claim number process if one does not believe in one particular Poisson process as claim arrival generating process. As a matter of fact, if we observed only one sample path $\widetilde{N}(\theta(\omega)\mu(t), \omega)$ of a mixed Poisson process, we would not be able to distinguish between this kind of process and a Poisson process with mean value function $\theta(\omega)\mu$. However, if we had several such sample paths we should see differences in the variation of the paths; see Figure 2.3.2 for an illustration of this phenomenon.

A mixed Poisson process is a special *Cox process* where the mean value function μ is a general random process with non-decreasing sample paths, independent of the underlying homogeneous Poisson process $\widetilde{N}$. Such processes have proved useful, for example, in medical statistics where every sample path represents the medical history of a particular patient which has his/her "own" mean value function. We can think of such a function as "drawn" from a distribution of mean value functions. Similarly, we can think of θ representing different factors of influence on an insurance portfolio. For example, think of the claim number process of a portfolio of car insurance policies as a collection of individual sample paths corresponding to the different insured persons. The variable $\theta(\omega)$ then represents properties such as the driving skill, the age, the driving experience, the health state, etc., of the individual drivers.

In Figure 2.3.2 we see one striking difference between a mixed Poisson process and a homogeneous Poisson process: the shape and magnitude of the sample paths of the mixed Poisson process vary significantly. This property *cannot* be explained by the mean value function

$$EN(t) = E\widetilde{N}(\theta\,\mu(t)) = E\big(E[\widetilde{N}(\theta\,\mu(t)) \mid \theta]\big) = E[\theta\,\mu(t)] = E\theta\,\mu(t)\,, \quad t \geq 0\,.$$

Thus, if $E\theta = 1$, as in Figure 2.3.2, the mean values of the random variables $\widetilde{N}(\mu(t))$ and $N(t)$ are the same. The differences between a mixed Poisson and a Poisson process with the same mean value function can be seen in the variances. First observe that the Poisson property implies

$$E(N(t) \mid \theta) = \theta\,\mu(t) \quad \text{and} \quad \text{var}(N(t) \mid \theta) = \theta\,\mu(t)\,. \tag{2.3.49}$$

Next we give an auxiliary result. Its prove is left as an exercise.

Lemma 2.3.4 *Let $A(B)$ be a measurable function of a random variable B such that* $\text{var}(A(B)) < \infty$. *Then*

$$\text{var}(A(B)) = E[\text{var}(A(B) \mid B)] + \text{var}(E[A(B) \mid B])\,.$$

An application of this formula with $A = N(t) = \widetilde{N}(\theta\mu(t))$ and $B = \theta$ together with (2.3.49) yields

$$\text{var}(N(t)) = E[\text{var}(N(t) \mid \theta)] + \text{var}(E[N(t) \mid \theta])$$

$$= E[\theta \, \mu(t)] + \text{var}(\theta \, \mu(t))$$

$$= E\theta \, \mu(t) + \text{var}(\theta) \, (\mu(t))^2$$

$$= EN(t) \left(1 + \frac{\text{var}(\theta)}{E\theta} \, \mu(t) \right)$$

$$> EN(t) \,,$$

where we assumed that $\text{var}(\theta) < \infty$ and $\mu(t) > 0$. The property

$$\text{var}(N(t)) > EN(t) \quad \text{for any } t > 0 \text{ with } \mu(t) > 0 \qquad (2.3.50)$$

is called *over-dispersion*. It is one of the major differences between a mixed Poisson process and a Poisson process N, where $EN(t) = \text{var}(N(t))$.

We conclude by summarizing some of the important properties of the mixed Poisson process; some of the proofs are left as exercises.

The mixed Poisson process *inherits* the following properties of the Poisson process:

- It has the *Markov property*; see Section 2.1.2 for some explanation.
- It has the *order statistics property*: if the function μ has a continuous a.e. positive intensity function λ and N has arrival times $0 < T_1 < T_2 < \cdots$, then for every $t > 0$,

$$(T_1, \ldots, T_n \mid N(t) = n) \stackrel{d}{=} (X_{(1)}, \ldots, X_{(n)}) \,,$$

where the right-hand side is the ordered sample of the iid random variables $X_1, \ldots, X_n$ with common density $\lambda(x)/\mu(t)$, $0 \le x \le t$; cf. Theorem 2.1.11.

The order statistics property is remarkable insofar that it does not depend on the mixing variable θ. In particular, for a mixed homogeneous Poisson process the conditional distribution of $(T_1, \ldots, T_{N(t)})$ given $\{N(t) = n\}$ is the distribution of the ordered sample of iid $U(0, t)$ distributed random variables.

The mixed Poisson process *loses* some of the properties of the Poisson process:

- It has *dependent increments*.
- In general, the distribution of $N(t)$ is *not Poisson*.
- It is *over-dispersed*; see (2.3.50).

Comments

For an extensive treatment of mixed Poisson processes and their properties we refer to the monograph by Grandell [37]. It can be shown that the mixed Poisson process and the Poisson process are the only *point processes* on $[0, \infty)$ which have the order statistics property; see Kallenberg [47]; cf. Grandell [37], Theorem 6.6.

Exercises

(1) Consider the mixed Poisson process $(N(t))_{t\geq 0} = (\tilde{N}(\theta t))_{t\geq 0}$ with arrival times T_i, where $\tilde{N}$ is a standard homogeneous Poisson process on $[0,\infty)$ and $\theta > 0$ is a non-degenerate mixing variable with $\text{var}(\theta) < \infty$, independent of $\tilde{N}$.

(a) Show that N does not have independent increments. (An easy way of doing this would be to calculate the covariance of $N(s,t]$ and $N(x,y]$ for disjoint intervals $(s,t]$ and $(x,y]$.)

(b) Show that N has the order statistics property, i.e., given $N(t) = n$, $(T_1,\ldots,T_n)$ has the same distribution as the ordered sample of the iid $U(0,t)$ distributed random variables $U_1,\ldots,U_n$.

(c) Calculate $P(N(t) = n)$ for $n \in \mathbb{N}_0$. Show that $N(t)$ is not Poisson distributed.

(d) The negative binomial distribution on $\{0,1,2,\ldots\}$ has the individual probabilities

$$p_k = \binom{v+k-1}{k} p^v (1-p)^k, \quad k \in \mathbb{N}_0, \quad p \in (0,1), \quad v > 0.$$

Consider the mixed Poisson process N with gamma distributed mixing variable, i.e., θ has $\Gamma(\gamma,\beta)$ density

$$f_\theta(x) = \frac{\beta^\gamma}{\Gamma(\gamma)} x^{\gamma-1} e^{-\beta x}, \quad x > 0.$$

Calculate the probabilities $P(N(t) = k)$ and give some reason why the process N is called *negative binomial process*.

(2) Give an algorithm for simulating the sample paths of an arbitrary mixed Poisson process.

(3) Prove Lemma 2.3.4.

(4) Let $N(t) = \tilde{N}(\theta t)$, $t \geq 0$, be mixed Poisson, where $\tilde{N}$ is a standard homogeneous Poisson process, independent of the mixing variable θ.

(a) Show that N satisfies the strong law of large numbers with random limit θ:

$$\frac{N(t)}{t} \to \theta \quad \text{a.s.}$$

(b) Show the following "central limit theorem":

$$\frac{N(t) - \theta t}{\sqrt{\theta t}} \xrightarrow{d} Y \sim N(0,1).$$

(c) Show that the "naive" central limit theorem does not hold by showing that

$$\frac{N(t) - EN(t)}{\sqrt{\text{var}(N(t))}} \xrightarrow{\text{a.s.}} \frac{\theta - E\theta}{\sqrt{\text{var}(\theta)}}.$$

Here we assume that $\text{var}(\theta) < \infty$.

(5) Let $N(t) = \tilde{N}(\theta t)$, $t \geq 0$, be mixed Poisson, where $\tilde{N}$ is a standard homogeneous Poisson process, independent of the mixing variable $\theta > 0$. Write F_θ for the distribution function of θ and $\overline{F}_\theta = 1 - F_\theta$ for its right tail. Show that the following relations hold for integer $n \geq 1$,

$$P(N(t) > n) = t \int_0^\infty \frac{(t\,x)^n}{n!}\,e^{-t\,x}\,\overline{F}_\theta(x)\,dx$$

$$P(\theta \le x \mid N(t) = n) = \frac{\int_0^x y^n\,e^{-y\,t}\,dF_\theta(y)}{\int_0^\infty y^n\,e^{-y\,t}\,dF_\theta(y)}\,,$$

$$E(\theta \mid N(t) = n) = \frac{\int_0^\infty y^{n+1}\,e^{-y\,t}\,dF_\theta(y)}{\int_0^\infty y^n\,e^{-y\,t}\,dF_\theta(y)}\,.$$

3

The Total Claim Amount

In Chapter 2 we learned about three of the most prominent claim number processes, N: the *Poisson process* in Section 2.1, the *renewal process* in Section 2.2, and the *mixed Poisson process* in Section 2.3. In this section we take a closer look at the *total claim amount process*, as introduced on p. 8:

$$S(t) = \sum_{i=1}^{N(t)} X_i, \quad t \geq 0, \tag{3.0.1}$$

where the *claim number process* N is independent of the iid *claim size sequence* (X_i). We also assume that $X_i > 0$ a.s. Depending on the choice of the process N, we get different models for the process S. In Example 2.1.3 we introduced the *Cramér-Lundberg model* as that particular case of model (3.0.1) when N is a homogeneous Poisson process. Another prominent model for S is called *renewal* or *Sparre-Anderson model*; it is model (3.0.1) when N is a renewal process.

In Section 3.1 we study the order of magnitude of the total claim amount $S(t)$ in the renewal model. This means we calculate the mean and the variance of $S(t)$ for large t, which give us a rough impression of the growth of $S(t)$ as $t \to \infty$. We also indicate that S satisfies the strong law of large numbers and the central limit theorem. The information about the asymptotic growth of the total claim amount enables one to give advise as to how much premium should be charged in a given time period in order to avoid bankruptcy or ruin in the portfolio. In Section 3.1.3 we collect some of the classical *premium calculation principles* which can be used as a rule of thumb for determining how big the premium income in a homogeneous portfolio should be.

We continue in Section 3.2 by considering some realistic claim size distributions and their properties. We consider exploratory statistical tools (QQ-plots, mean excess function) and apply them to real-life claim size data in order to get a preliminary understanding of which distributions fit real-life data. In this context, the issue of modeling large claims deserves particular attention. We discuss the notions of heavy- and light-tailed claim size distributions as appropriate for modeling large and small claims, respectively. Then, in Sections 3.2.5 and 3.2.6 we focus on the subexponential distributions and on distributions with regularly varying tails. The latter classes contain those distributions which are most appropriate for modeling large claims.

In Section 3.3 we study finally the distribution of the total claim amount $S(t)$ as a combination of claim number process and claim sizes. We start in Section 3.3.1 by investigating some theoretical properties of the total claim amount models. By applying characteristic function techniques, we learn about *mixture distributions* as useful tools in the context of compound Poisson and compound geometric processes. We show that the summation of independent compound Poisson processes yields a compound Poisson process and we investigate consequences of this result. In particular, we show in the framework of the Cramér-Lundberg model that the total claim amounts from disjoint layers for the claim sizes or over disjoint periods of time are independent compound Poisson variables. We continue in Section 3.3.3 with a numerical recursive procedure for determining the distribution of the total claim amount. In the insurance world, this technique is called *Panjer recursion*. In Sections 3.3.4 and 3.3.5 we consider alternative methods for determining *approximations* to the distribution of the total claim amount. These approximations are based on the central limit theorem or Monte Carlo techniques.

Finally, in Section 3.4 we apply the developed theory to the case of reinsurance treaties. The latter are agreements between a primary and a secondary insurer with the aim to protect the primary insurer against excessive losses which are caused by very large claim sizes or by a large number of small and moderate claim sizes. We discuss the most important forms of the treaties and indicate how previously developed theory can be applied to deal with their distributional properties.

3.1 The Order of Magnitude of the Total Claim Amount

Given a particular model for S, one of the important questions for an insurance company is to determine the order of magnitude of $S(t)$. This information is needed in order to determine a *premium* which covers the *losses* represented by $S(t)$.

Most desirably, one would like to know the distribution of $S(t)$. This, however, is in general a too complicated problem and therefore one often relies on numerical or simulation methods in order to approximate the distribution of $S(t)$. In this section we consider some simple means in order to get a rough impression of the size of the total claim amount. Those means include the expectation and variance of $S(t)$ (Section 3.1.1), the strong law of large numbers, and the central limit theorem for $S(t)$ as $t \to \infty$ (Section 3.1.2). In Section 3.1.3 we study the relationship of these results with premium calculation principles.

3.1.1 The Mean and the Variance in the Renewal Model

The expectation of a random variable tells one about its average size. For the total claim amount the expectation is easily calculated by exploiting the independence of (X_i) and $N(t)$, provided $EN(t)$ and EX_1 are finite:

$$ES(t) = E\left[E\left(\sum_{i=1}^{N(t)} X_i \;\middle|\; N(t)\right)\right] = E\left(N(t)\,EX_1\right) = EN(t)\,EX_1\,.$$

Example 3.1.1 (Expectation of $S(t)$ in the Cramér-Lundberg and renewal models)
In the Cramér-Lundberg model, $EN(t) = \lambda t$, where λ is the intensity of the homogeneous Poisson process N. Hence

$$ES(t) = \lambda t\, EX_1\,.$$

Such a compact formula does not exist in the general renewal model. However, given $EW_1 = \lambda^{-1} < \infty$ we know from the elementary renewal Theorem 2.2.7 that $EN(t)/t \to \lambda$ a.s. as $t \to \infty$. Therefore

$$ES(t) = \lambda t\, EX_1\,(1 + o(1))\,, \quad t \to \infty\,.$$

This is less precise information than in the Cramér-Lundberg model. However, this formula tells us that the expected total claim amount grows roughly linearly for large t. As in the Cramér-Lundberg case, the slope of the linear function is determined by the reciprocal of the expected inter-arrival time EW_1 and the expected claim size EX_1. □

The expectation does not tell one too much about the distribution of $S(t)$. We learn more about the order of magnitude of $S(t)$ if we combine the information about $ES(t)$ with the variance $\mathrm{var}(S(t))$.

Assume that $\mathrm{var}(N(t))$ and $\mathrm{var}(X_1)$ are finite. Conditioning on $N(t)$ and exploiting the independence of $N(t)$ and (X_i), we obtain

$$\mathrm{var}\left[\sum_{i=1}^{N(t)}(X_i - EX_1)\;\middle|\; N(t)\right] = \sum_{i=1}^{N(t)} \mathrm{var}(X_i \mid N(t))$$

$$= N(t)\,\mathrm{var}(X_1 \mid N(t)) = N(t)\,\mathrm{var}(X_1)$$

$$E\left[\sum_{i=1}^{N(t)} X_i \;\middle|\; N(t)\right] = N(t)\,EX_1\,.$$

By virtue of Lemma 2.3.4 we conclude that

$$\mathrm{var}(S(t)) = E[N(t)\,\mathrm{var}(X_1)] + \mathrm{var}(N(t)\,EX_1)$$

$$= EN(t)\,\mathrm{var}(X_1) + \mathrm{var}(N(t))\,(EX_1)^2\,.$$

Example 3.1.2 (Variance of $S(t)$ in the Cramér-Lundberg and renewal models)

In the Cramér-Lundberg model the Poisson distribution of $N(t)$ gives us $EN(t) = \text{var}(N(t)) = \lambda t$. Hence

$$\text{var}(S(t)) = \lambda t \left[\text{var}(X_1) + (EX_1)^2\right] = \lambda t E(X_1^2).$$

In the renewal model we again depend on some asymptotic formulae for $EN(t)$ and $\text{var}(N(t))$; see Theorem 2.2.7 and Proposition 2.2.10:

$$\text{var}(S(t)) = \left[\lambda t \,\text{var}(X_1) + \text{var}(W_1) \lambda^3 t \,(EX_1)^2\right] (1 + o(1))$$

$$= \lambda t \left[\text{var}(X_1) + \text{var}(W_1) \lambda^2 \,(EX_1)^2\right] (1 + o(1)).$$

$\square$

We summarize our findings.

Proposition 3.1.3 (Expectation and variance of the total claim amount in the renewal model)
In the renewal model, if $EW_1 = \lambda^{-1}$ and EX_1 are finite,

$$\lim_{t \to \infty} \frac{ES(t)}{t} = \lambda EX_1,$$

and if $\text{var}(W_1)$ and $\text{var}(X_1)$ are finite,

$$\lim_{t \to \infty} \frac{\text{var}(S(t))}{t} = \lambda \left[\text{var}(X_1) + \text{var}(W_1) \lambda^2 \,(EX_1)^2\right].$$

In the Cramér-Lundberg model these limit relations degenerate to identities for every $t > 0$:

$$ES(t) = \lambda t EX_1 \quad and \quad \text{var}(S(t)) = \lambda t E(X_1^2).$$

The message of these results is that in the renewal model *both the expectation and the variance of the total claim amount grow roughly linearly as a function of t.* This is important information which can be used to give a rule of thumb about how much premium has to be charged for covering the losses $S(t)$: the premium should increase roughly linearly and with a slope larger than λEX_1. In Section 3.1.3 we will consider some of the classical premium calculation principles and there we will see that this rule of thumb is indeed quite valuable.

3.1.2 The Asymptotic Behavior in the Renewal Model

In this section we are interested in the asymptotic behavior of the total claim amount process. Throughout we assume the renewal model (see p. 77) for the total claim amount process S. As a matter of fact, $S(t)$ satisfies quite a general strong law of large numbers and central limit theorem:

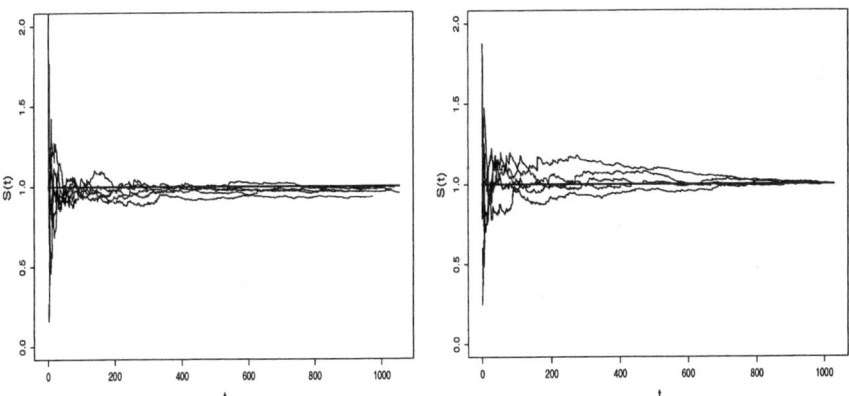

Figure 3.1.4 *Visualization of the strong law of large numbers for the total claim amount S in the Cramér-Lundberg model with unit Poisson intensity. Five sample paths of the process* $(S(t)/t)$ *are drawn in the interval* $[0, 1000]$. *Left: Standard exponential claim sizes. Right: Pareto distributed claim sizes* $X_i = 1 + (Y_i - EY_1)/\sqrt{\mathrm{var}(Y_1)}$ *for iid* Y_i*'s with distribution function* $P(Y_i \leq x) = 1 - 2^4 x^{-4}$, $x \geq 2$. *These random variables have mean and variance 1. The fluctuations of* $S(t)/t$ *around the mean 1 for small t are more pronounced than for exponential claim sizes. The right tail of the distribution of* X_1 *is much heavier than the right tail of the exponential distribution. Therefore much larger claim sizes may occur.*

Theorem 3.1.5 (The strong law of large numbers and the central limit theorem in the renewal model)
Assume the renewal model for S.

(1) *If the inter-arrival times* W_i *and the claim sizes* X_i *have finite expectation, S satisfies the strong law of large numbers:*

$$\lim_{t \to \infty} \frac{S(t)}{t} = \lambda\, EX_1 \quad \text{a.s.} \tag{3.1.2}$$

(2) *If the inter-arrival times* W_i *and the claim sizes* X_i *have finite variance, S satisfies the central limit theorem:*

$$\sup_{x \in \mathbb{R}} \left| P\left(\frac{S(t) - ES(t)}{\sqrt{\mathrm{var}(S(t))}} \leq x \right) - \Phi(x) \right| \to 0, \tag{3.1.3}$$

where Φ *is the distribution function of the standard normal* $\mathrm{N}(0, 1)$ *distribution.*

Notice that the random sum process S satisfies essentially the same invariance principles, strong law of large numbers and central limit theorem, as the partial sum process

$$S_n = X_1 + \cdots + X_n, \quad n \geq 1.$$

Indeed, we know from a course in probability theory that (S_n) satisfies the strong law of large numbers

$$\lim_{n \to \infty} \frac{S_n}{n} = EX_1 \quad \text{a.s.,} \tag{3.1.4}$$

provided $EX_1 < \infty$, and the central limit theorem

$$P\left(\frac{S_n - ES_n}{\sqrt{\text{var}(S_n)}} \leq x\right) \to \Phi(x), \quad x \in \mathbb{R},$$

provided $\text{var}(X_1) < \infty$.

In both relations (3.1.2) and (3.1.3) we could use the asymptotic expressions for $ES(t)$ and $\text{var}(S(t))$ suggested in Proposition 3.1.3 for normalizing and centering purposes. Indeed, we have

$$\lim_{t \to \infty} \frac{S(t)}{ES(t)} = 1 \quad \text{a.s.}$$

and it can be shown by using some more sophisticated asymptotics for $ES(t)$ that as $t \to \infty$,

$$\sup_{x \in \mathbb{R}} \left| P\left(\frac{S(t) - \lambda\, EX_1\, t}{\sqrt{\lambda t\, [\text{var}(X_1) + \text{var}(W_1)\, \lambda^2\, (EX_1)^2]}} \leq x\right) - \Phi(x) \right| \to 0.$$

We also mention that the uniform version (3.1.3) of the central limit theorem is equivalent to the pointwise central limit theorem

$$P\left(\frac{S(t) - ES(t)}{\sqrt{\text{var}(S(t))}} \leq x\right) \to \Phi(x), \quad x \in \mathbb{R}.$$

This is a consequence of the well-known fact that convergence in distribution with continuous limit distribution function implies uniformity of this convergence; see Billingsley [13].

Proof. We only prove the first part of the theorem. For the second part, we refer to Embrechts et al. [29], Theorem 2.5.16. We have

$$\frac{S(t)}{t} = \frac{S(t)}{N(t)} \frac{N(t)}{t}. \tag{3.1.5}$$

Write

$$\Omega_1 = \{\omega : N(t)/t \to \lambda\} \quad \text{and} \quad \Omega_2 = \{\omega : S(t)/N(t) \to EX_1\}.$$

By virtue of (3.1.5) the result follows if we can show that $P(\Omega_1 \cap \Omega_2) = 1$. However, we know from the strong law of large numbers for N (Theorem 2.2.4)

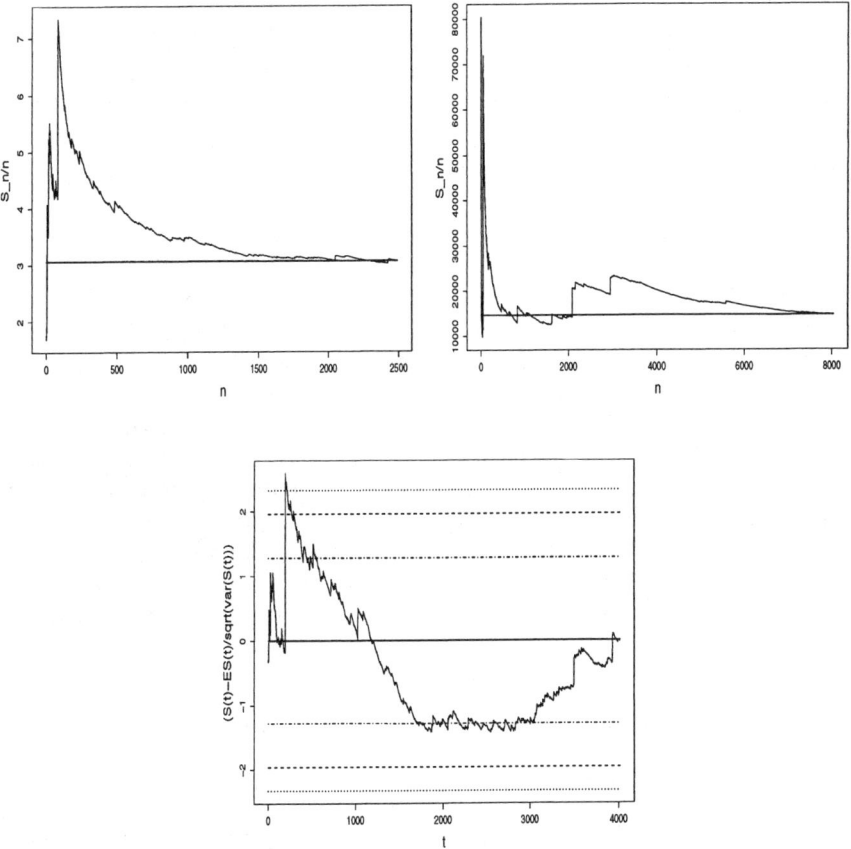

Figure 3.1.6 Top: *Visualization of the strong law of large numbers for the Danish fire insurance data* (left) *and the US industrial fire data* (right). *For a description of these data sets, see Example 3.2.11. The curves show the averaged sample sizes* $S_n/n = (X_1 + \cdots + X_n)/n$ *as a function of* n; *the solid straight line represents the overall sample mean. Both claim size samples contain very large values. This fact makes the ratio* S_n/n *converge to* EX_1 *very slowly. Bottom: The quantities* $(S(t) - ES(t))/\sqrt{\mathrm{var}(S(t))}$ *for the Danish fire insurance data. The values of* $ES(t)$ *and* $\mathrm{var}(S(t))$ *were evaluated from the asymptotic expressions suggested by Proposition 3.1.3. From bottom to top, the constant lines correspond to the 1%-, 2.5%-, 10%-, 50%-, 90%-, 97.5%-, 99%-quantiles of the standard normal distribution.*

that $P(\Omega_1) = 1$. Moreover, since $N(t) \overset{\text{a.s.}}{\to} \infty$, an application of the strong law of large numbers (3.1.4) and Lemma 2.2.6 imply that $P(\Omega_2) = 1$. This concludes the proof. □

The strong law of large numbers for the total claim amount process S is one of the important results which any insurance business has experienced since the foundation of insurance companies. As a matter of fact, the strong law of large numbers can be observed in real-life data; see Figure 3.1.6. Its validity gives one confidence that large and small claims averaged over time converge to their theoretical mean value. The strong law of large numbers and the central limit theorem for S are backbone results when it comes to premium calculation. This is the content of the next section.

3.1.3 Classical Premium Calculation Principles

One of the basic questions of an insurance business is how one chooses a premium in order to cover the losses over time, described by the total claim amount process S. We think of the premium income $p(t)$ in the portfolio of those policies where the claims occur as a deterministic function.

A coarse, but useful approximation to the random quantity $S(t)$ is given by its expectation $ES(t)$. Based on the results of Sections 3.1.1 and 3.1.2 for the renewal model, we would expect that the insurance company loses on average if $p(t) < ES(t)$ for large t and and gains if $p(t) > ES(t)$ for large t. Therefore it makes sense to choose a premium by "loading" the expected total claim amount by a certain positive number ρ.

For example, we know from Proposition 3.1.3 that in the renewal model

$$ES(t) = \lambda\, EX_1\, t\, (1 + o(1))\,, \quad t \to \infty\,.$$

Therefore it is reasonable to choose $p(t)$ according to the equation

$$p(t) = (1 + \rho)\, ES(t) \quad \text{or} \quad p(t) = (1 + \rho)\, \lambda\, EX_1\, t\,, \tag{3.1.6}$$

for some positive number ρ, called the *safety loading*. From the asymptotic results in Sections 3.1.1 and 3.1.2 it is evident that the insurance business is the more on the safe side the larger ρ. On the other hand, an overly large value ρ would make the insurance business less competitive: the number of contracts would decrease if the premium were too high compared to other premiums offered in the market. Since the success of the insurance business is based on the strong law of large numbers, one needs large numbers of policies in order to ensure the balance of premium income and total claim amount. Therefore, premium calculation principles more sophisticated than those suggested by (3.1.6) have also been considered in the literature. We briefly discuss some of them.

- The *net* or *equivalence principle*. This principle determines the premium $p(t)$ at time t as the expectation of the total claim amount $S(t)$:

$$p_{\text{Net}}(t) = ES(t)\,.$$

In a sense, this is the "fair market premium" to be charged: the insurance portfolio does not lose or gain capital on average. However, the central limit theorem (Theorem 3.1.3) in the renewal model tells us that the deviation of $S(t)$ from its mean increases at an order comparable to its standard deviation $\sqrt{\mathrm{var}(S(t))}$ as $t \to \infty$. Moreover, these deviations can be both positive or negative with positive probability. Therefore it would be utterly unwise to charge a premium according to this calculation principle. It is of purely theoretical value, a "benchmark premium". In Section 4.1 we will see that the net principle leads to "ruin" of the insurance business.

* The *expected value principle*.

$$p_{EV}(t) = (1 + \rho)\, ES(t)\,,$$

for some positive *safety loading* ρ. The rationale of this principle is the strong law of large numbers of Theorem 3.1.5, as explained above.

* The *variance principle*.

$$p_{Var}(t) = ES(t) + \alpha\, \mathrm{var}(S(t))\,,$$

for some positive α. In the renewal model, this principle is equivalent in an asymptotic sense to the expected value principle with a positive loading. Indeed, using Proposition 3.1.3, it is not difficult to see that the ratio of the premiums charged by both principles converges to a positive constant as $t \to \infty$, and α plays the role of a positive safety loading.

* The *standard deviation principle*.

$$p_{SD}(t) = ES(t) + \alpha\, \sqrt{\mathrm{var}(S(t))}\,,$$

for some positive α. The rationale for this principle is the central limit theorem since in the renewal model (see Theorem 3.1.5),

$$P(S(t) - p_{SD}(t) \le x) \to \Phi(\alpha)\,, \quad x \in \mathbb{R}\,,$$

where Φ is the standard normal distribution function. Convince yourself that this relation holds. In the renewal model, the standard deviation principle and the net principle are equivalent in the sense that the ratio of the two premiums converges to 1 as $t \to \infty$. This means that one charges a smaller premium by using this principle in comparison to the expected value and variance principles.

The interpretation of the premium calculation principles depends on the underlying model. In the renewal and Cramér-Lundberg models the interpretation follows by using the central limit theorem and the strong law of large numbers. If we assumed the mixed homogeneous Poisson process as the claim number process, the over-dispersion property, i.e., $\mathrm{var}(N(t)) > EN(t)$, would lead to completely different statements. For example, for a mixed compound homogeneous Poisson process $p_{Var}(t)/p_{EV}(t) \to \infty$ as $t \to \infty$. Verify this!

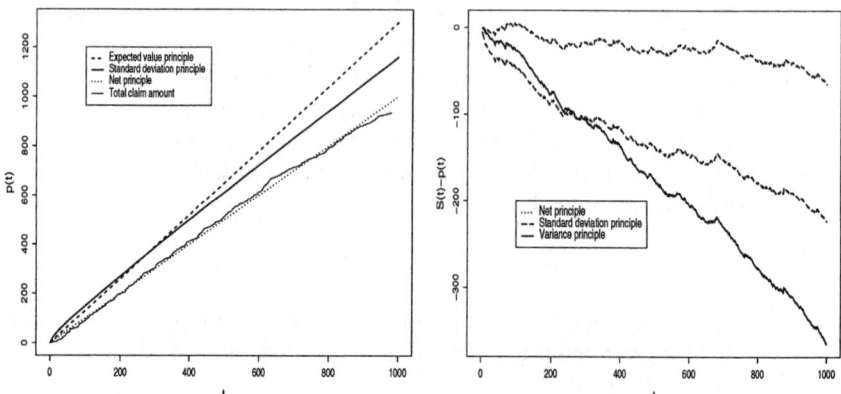

Figure 3.1.7 *Visualization of the premium calculation principles in the Cramér-Lundberg model with Poisson intensity 1 and standard exponential claim sizes. Left: The premiums are: for the net principle $p_{\mathrm{Net}}(t) = t$, for the standard deviation principle $p_{\mathrm{SD}}(t) = t + 5\sqrt{2t}$ and for the expected value principle $p_{\mathrm{EV}}(t) = 1.3t$ for $\rho = 0.3$. Equivalently, $p_{\mathrm{EV}}(t)$ corresponds to the variance principle $p_{\mathrm{Var}}(t) = 1.3t$ with $\alpha = 0.15$. One sample path of the total claim amount process S is also given. Notice that $S(t)$ can lie above or below $p_{\mathrm{Net}}(t)$. Right: The differences $S(t) - p(t)$ are given. The upper curve corresponds to p_{Net}.*

Comments

Various other theoretical premium principles have been introduced in the literature; see for example Bühlmann [19], Kaas et al. [46] or Klugman et al. [51]. In Exercise 2 below one finds theoretical requirements taken from the actuarial literature that a "reasonable" premium calculation principle should satisfy. As a matter of fact, just one of these premium principles satisfies all requirements. It is the net premium principle which is not reasonable from an economic point of view since its application leads to ruin in the portfolio.

Exercises

(1) Assume the renewal model for the total claim amount process S with $\mathrm{var}(X_1) < \infty$ and $\mathrm{var}(W_1) < \infty$.

(a) Show that the standard deviation principle is motivated by the central limit theorem, i.e., as $t \to \infty$,

$$P(S(t) - p_{\mathrm{SD}}(t) \le x) \to \Phi(\alpha), \quad x \in \mathbb{R},$$

where Φ is the standard normal distribution. This means that α is the $\Phi(\alpha)$-quantile of the normal distribution.

(b) Show that the net principle and the standard deviation principle are asymptotically equivalent in the sense that

$$\frac{p_{\text{Net}}(t)}{p_{\text{SD}}(t)} \to 1 \quad \text{as } t \to \infty.$$

(c) Argue why the net premium principle and the standard deviation principle are "sufficient for a risk neutral insurer only", i.e., these principles do not lead to a positive relative average profit in the long run: consider the relative gains $(p(t) - ES(t))/ES(t)$ for large t.

(d) Show that for $h > 0$,

$$\lim_{t \to \infty} \frac{ES(t - h, t]}{t} = h \frac{EX_1}{EW_1}.$$

Hint: Appeal to Blackwell's renewal theorem; see p. 66.

(2) In the insurance literature one often finds theoretical requirements on the premium principles. Here are a few of them:

- *Non-negative loading* : $p(t) \geq ES(t)$.
- *Consistency* : the premium for $S(t) + c$ is $p(t) + c$.
- *Additivity* : for independent total claim amounts $S(t)$ and $S'(t)$ with corresponding premiums $p(t)$ and $p'(t)$, the premium for $S(t) + S'(t)$ should be $p(t) + p'(t)$.
- *Homogeneity* or *proportionality* : for $c > 0$, the premium for $c S(t)$ should be $c p(t)$.

Which of the premium principles satisfies these conditions in the Cramér-Lundberg or renewal models?

(3) Calculate the mean and the variance of the total claim amount $S(t)$ under the condition that N is mixed Poisson with $(N(t))_{t \geq 0} = (\widetilde{N}(\theta\, t))_{t \geq 0}$, where $\widetilde{N}$ is a standard homogeneous Poisson process, $\theta > 0$ is a mixing variable with $\text{var}(\theta) < \infty$, and (X_i) is an iid claim size sequence with $\text{var}(X_1) < \infty$. Show that

$$p_{\text{Var}}(t)/p_{\text{EV}}(t) \to \infty, \quad t \to \infty.$$

Compare the latter limit relation with the case when N is a renewal process.

(4) Assume the Cramér-Lundberg model with Poisson intensity $\lambda > 0$ and consider the corresponding risk process

$$U(t) = u + ct - S(t),$$

where $u > 0$ is the initial capital in the portfolio, $c > 0$ the premium rate and S the total claim amount process. The risk process and its meaning are discussed in detail in Chapter 4. In addition, assume that the moment generating function $m_{X_1}(h) = E \exp\{h X_1\}$ of the claim sizes X_i is finite in some neighborhood $(-h_0, h_0)$ of the origin.

(a) Calculate the moment generating function of $S(t)$ and show that it exists in $(-h_0, h_0)$.

(b) The premium rate c is determined according to the expected value principle: $c = (1 + \rho)\, \lambda\, EX_1$ for some positive safety loading ρ, where the value c (equivalently, the value ρ) can be chosen according to the *exponential premium principle*.[1] For its definition, write $v_\alpha(u) = e^{-\alpha u}$ for $u, \alpha > 0$. Then c is chosen as the solution to the equation

$$v_\alpha(u) = E[v_\alpha(U(t))] \quad \text{for all } t > 0. \tag{3.1.7}$$

Use (a) to show that a unique solution $c = c_\alpha > 0$ to (3.1.7) exists. Calculate the safety loading ρ_α corresponding to c_α and show that $\rho_\alpha \geq 0$.

(c) Consider c_α as a function of $\alpha > 0$. Show that $\lim_{\alpha \downarrow 0} c_\alpha = \lambda E X_1$. This means that c_α converges to the value suggested by the net premium principle with safety loading $\rho = 0$.

3.2 Claim Size Distributions

In this section we are interested in the question:

What are realistic claim size distributions?

This question is about the goodness of fit of the claim size data to the chosen distribution. It is not our goal to give sophisticated statistical analyzes, but we rather aim at introducing some classes of distributions used in insurance practice, which are sufficiently flexible and give a satisfactory fit to the data. In Section 3.2.1 we introduce QQ-plots and in Section 3.2.3 mean excess plots as two graphical methods for discriminating between different claim size distributions. Since realistic claim size distributions are very often heavy-tailed, we start in Section 3.2.2 with an informal discussion of the notions of heavy- and light-tailed distributions. In Section 3.2.4 we introduce some of the major claim size distributions and discuss their properties. In Sections 3.2.5 and 3.2.6 we continue to discuss natural heavy-tailed distributions for insurance: the classes of the distributions with regularly varying tails and the subexponential distributions. The latter class is by now considered as *the* class of distributions for modeling large claims.

3.2.1 An Exploratory Statistical Analysis: QQ-Plots

We consider some simple exploratory statistical tools and apply them to simulated and real-life claim size data in order to detect which distributions might give a reasonable fit to real-life insurance data. We start with a *quantile-quantile plot*, for short *QQ-plot*, and continue in Section 3.2.3 with a mean excess plot. Quantiles correspond to the "inverse" of a distribution function, which is not always well-defined (distribution functions are not necessarily strictly increasing). We focus on a left-continuous version.

---

[1] This premium calculation principle is not intuitively motivated by the strong law of large numbers or the central limit theorem, but by so-called *utility theory*. The reader who wants to learn about the rationale of this principle is referred to Chapter 1 in Kaas et al. [46].

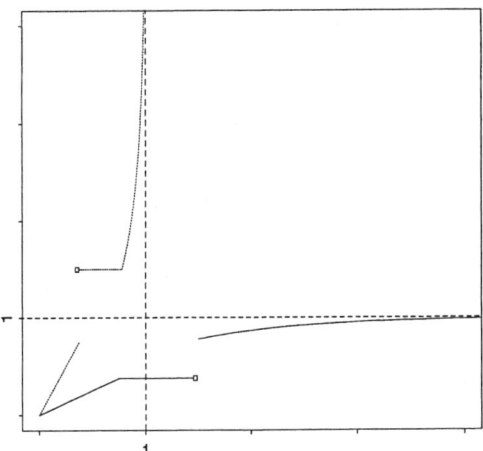

▷**Figure 3.2.1** *A distribution function F on $[0, \infty)$ and its quantile function $F^{\leftarrow}$. In a sense, $F^{\leftarrow}$ is the mirror image of F with respect to the line $x = y$.*

Definition 3.2.2 (Quantile function)
The generalized inverse of the distribution function F, i.e.,

$$F^{\leftarrow}(t) = \inf\{x \in \mathbb{R} : F(x) \geq t\}, \quad 0 < t < 1,$$

is called the quantile function *of the distribution function F. The quantity $x_t = F^{\leftarrow}(t)$ defines the t-quantile of F.*

If F is monotone increasing (such as the distribution function Φ of the standard normal distribution), we see that $F^{\leftarrow} = F^{-1}$, i.e., the ordinary inverse of F. An illustration of the quantile function is given in Figure 3.2.1. Notice that intervals where F is constant turn into jumps of $F^{\leftarrow}$, and jumps of F turn into intervals of constancy for $F^{\leftarrow}$.

In this way we can define the generalized inverse of the *empirical distribution function* F_n of a sample $X_1, \ldots, X_n$, i.e.,

$$F_n(x) = \frac{1}{n} \sum_{i=1}^{n} I_{(-\infty, x]}(X_i), \quad x \in \mathbb{R}. \tag{3.2.8}$$

It is easy to verify that F_n has all properties of a distribution function:

- $\lim_{x \to -\infty} F_n(x) = 0$ and $\lim_{x \to \infty} F_n(x) = 1$.
- F_n is non-decreasing: $F_n(x) \leq F_n(y)$ for $x \leq y$.
- F_n is right-continuous: $\lim_{y \downarrow x} F_n(y) = F_n(x)$ for every $x \in \mathbb{R}$.

Let $X_{(1)} \leq \cdots \leq X_{(n)}$ be the ordered sample of $X_1, \ldots, X_n$. In what follows, we assume that the sample does not have ties, i.e., $X_{(1)} < \cdots < X_{(n)}$ a.s. For example, if the X_i's are iid with a density the sample does not have ties; see the proof of Lemma 2.1.9 for an argument.

Since the empirical distribution function of a sample is itself a distribution function, one can calculate its quantile function $F_n^{\leftarrow}$ which we call the *empirical quantile function*. If the sample has no ties then it is not difficult to see that

$$F_n(X_{(k)}) = k/n, \quad k = 1, \ldots, n,$$

i.e., F_n jumps by $1/n$ at every value $X_{(k)}$ and is constant in $[X_{(k)}, X_{(k+1)})$ for $k < n$. This means that the empirical quantile function $F_n^{\leftarrow}$ jumps at the values k/n by $X_{(k)} - X_{(k-1)}$ and remains constant in $((k-1)/n, k/n]$:

$$F_n^{\leftarrow}(t) = \begin{cases} X_{(k)} & t \in ((k-1)/n, k/n], \quad k = 1, \ldots, n-1, \\ X_{(n)} & t \in ((n-1)/n, 1). \end{cases}$$

A fundamental result of probability theory, the *Glivenko-Cantelli lemma*, (see for example Billingsley [13], p. 275) tells us the following: if $X_1, X_2, \ldots$ is an iid sequence with distribution function F, then

$$\sup_{x \in \mathbb{R}} |F_n(x) - F(x)| \overset{a.s.}{\to} 0,$$

implying that $F_n(x) \approx F(x)$ uniformly for all x. One can show that the Glivenko-Cantelli lemma implies $F_n^{\leftarrow}(t) \to F^{\leftarrow}(t)$ a.s. as $n \to \infty$ for all continuity points t of $F^{\leftarrow}$; see Resnick [64], p. 5. This observation is the basic idea for the *QQ-plot*: if $X_1, \ldots, X_n$ were a sample with known distribution function F, we would expect that $F_n^{\leftarrow}(t)$ is close to $F^{\leftarrow}(t)$ for all $t \in (0,1)$, provided n is large. Thus, if we plot $F_n^{\leftarrow}(t)$ against $F^{\leftarrow}(t)$ for $t \in (0,1)$ we should roughly see a straight line.

It is common to plot the graph

$$\left\{ \left(X_{(k)}, F^{\leftarrow}\left(\frac{k}{n+1} \right) \right), \quad k = 1, \ldots, n \right\}$$

for a given distribution function F. Modifications of the plotting positions have been used as well. Chambers [21] gives the following properties of a QQ-plot:

(a) Comparison of distributions. *If the data were generated from a random sample of the reference distribution, the plot should look roughly linear. This remains true if the data come from a linear transformation of the distribution.*

(b) Outliers. *If one or a few of the data values are contaminated by gross error or for any reason are markedly different in value from the remaining values, the latter being more or less distributed like the reference distribution, the outlying points may be easily identified on the plot.*

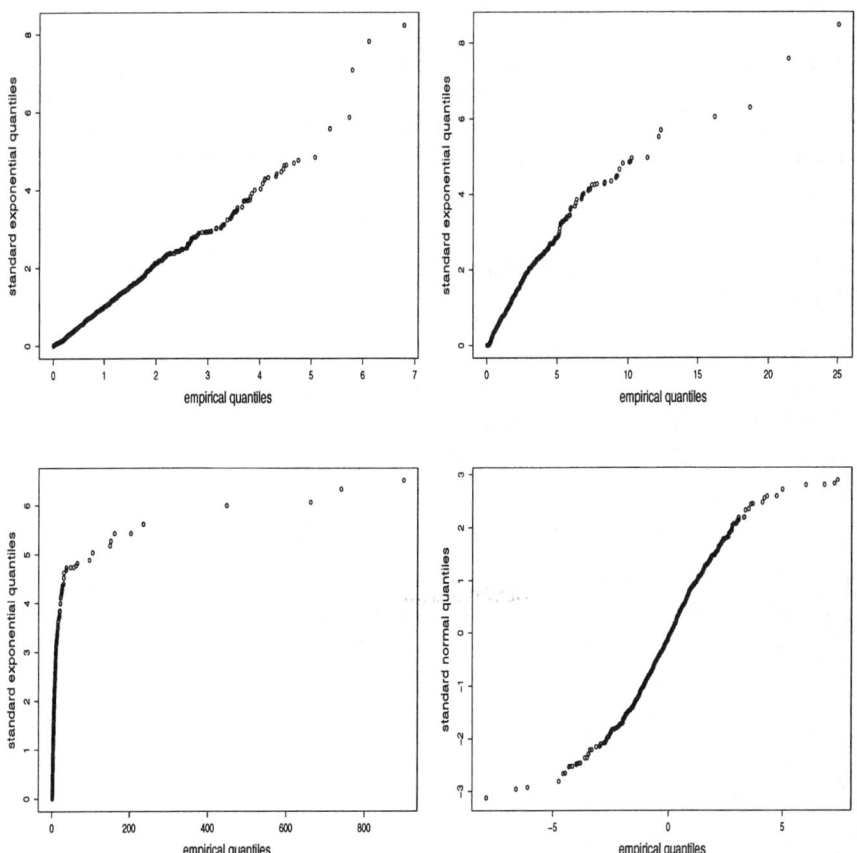

Figure 3.2.3 *QQ-plots for samples of size* 1 000. *Standard exponential* (top left), *standard log-normal* (top right) *and Pareto distributed data with tail index* 4 (bottom left) *versus the standard exponential quantiles.* Bottom right: *t_4-distributed data versus the quantiles of the standard normal distribution. The t_4-distribution has tails $F(-x) = 1 - F(x) = c\,x^{-4}(1 + o(1))$ as $x \to \infty$, some $c > 0$, in contrast to the standard normal with tails $\Phi(-x) = 1 - \Phi(x) = (\sqrt{2\pi}x)^{-1}\exp\{-x^2/2\}(1 + o(1))$; see* (3.2.9).

(c) Location and scale. *Because a change of one of the distributions by a linear transformation* simply transforms the plot by the same transformation, *one may estimate graphically (through the intercept and slope) location and scale parameters for a sample of data, on the assumption that the data come from the reference distribution.*

(d) Shape. *Some difference in distributional shape may be deduced from the plot. For example if the reference distribution has heavier tails (tends to have more large values) the plot will curve down at the left and/or up at the right.*

For an illustration of (a) and (d), also for a two-sided distribution, see Figure 3.2.3. QQ-plots applied to real-life claim size data (Danish fire insurance, US industrial fire) are presented in Figures 3.2.5 and 3.2.15. QQ-plots applied to the Danish fire insurance inter-arrival times are given in Figures 2.1.22 and 2.1.23.

3.2.2 A Preliminary Discussion of Heavy- and Light-Tailed Distributions

The Danish fire insurance data and the US industrial fire data presented in Figures 3.2.5 and 3.2.15, respectively, can be modeled by a very heavy-tailed distribution. Such claim size distributions typically occur in a reinsurance portfolio, where the largest claims are insured. In this context, the question arises:

What determines a heavy-tailed/light-tailed claim size distribution?

There is no clear-cut answer to this question. One common way to characterize the heaviness of the tails is by means of the exponential distribution as a benchmark. For example, if

$$\limsup_{x\to\infty} \frac{\overline{F}(x)}{e^{-\lambda x}} < \infty \quad \text{for some } \lambda > 0,$$

where

$$\overline{F}(x) = 1 - F(x), \quad x > 0,$$

denotes the right tail of the distribution function F, we could call F *light-tailed*, and if

$$\liminf_{x\to\infty} \frac{\overline{F}(x)}{e^{-\lambda x}} > 0 \quad \text{for all } \lambda > 0,$$

we could call F *heavy-tailed*.

Example 3.2.4 (Some well-known heavy- and light-tailed claim size distributions)
From the above definitions, the exponential $\text{Exp}(\lambda)$ distribution is light-tailed for every $\lambda > 0$.

A standard claim size distribution is the *truncated normal*. This means that the X_i's have distribution function $F(x) = P(|Y| \leq x)$ for a normally distributed random variable Y. If we assume Y standard normal, $F(x) = 2(\Phi(x) - 0.5)$ for $x > 0$, where Φ is the standard normal distribution function with density

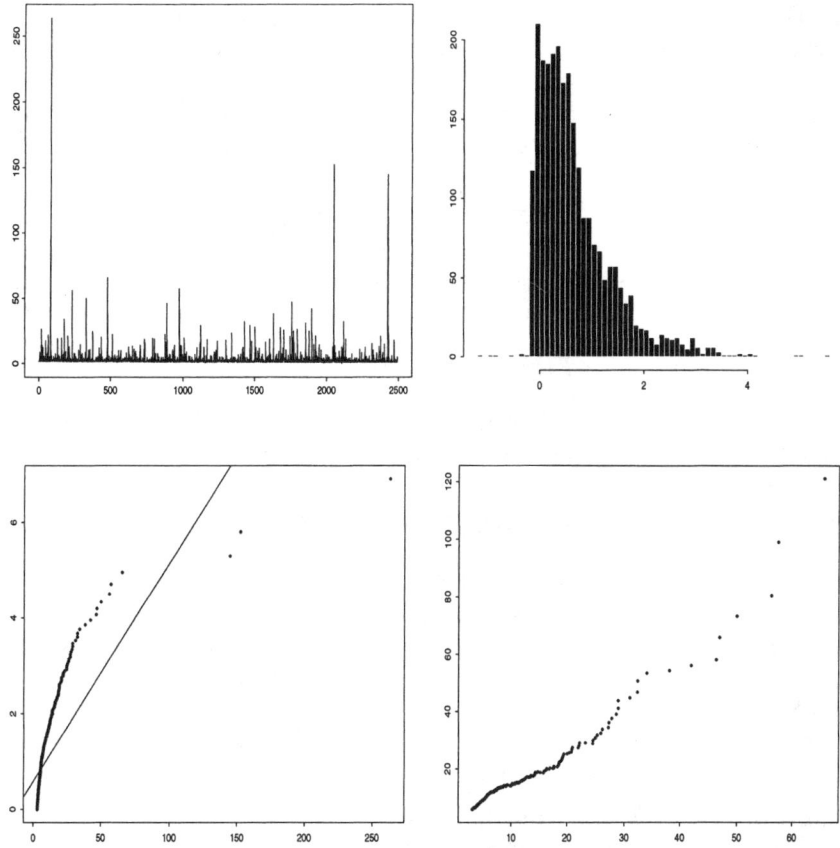

Figure 3.2.5 Top left: *Danish fire insurance claim size data in millions of Danish Kroner* (1985 *prices*). *The data correspond to the period* 1980 − 1992. *There is a total of* 2 493 *observations.* Top right: *Histogram of the log-data.* Bottom left: *QQ-plot of the data against the standard exponential distribution. The graph is curved down at the right indicating that the right tail of the distribution of the data is significantly heavier than the exponential.* Bottom right: *Mean excess plot of the data. The graph increases in its whole domain. This is a strong indication of heavy tails of the underlying distribution. See Example 3.2.11 for some comments.*

$$\varphi(x) = \frac{\mathrm{e}^{-x^2/2}}{\sqrt{2\pi}}, \quad x \in \mathbb{R}.$$

An application of l'Hospital's rule shows that

$$\lim_{x \to \infty} \frac{\overline{\Phi}(x)}{x^{-1}\varphi(x)} = 1. \tag{3.2.9}$$

The latter relation is often referred to as *Mill's ratio*. With Mill's ratio in mind, it is easy to verify that the truncated normal distribution is light-tailed. Using an analogous argument, it can be shown that the gamma distribution, for any choice of parameters, is light-tailed. Verify this.

A typical example of a heavy-tailed claim size distribution is the *Pareto distribution* with tail parameter $\alpha > 0$ and scale parameter $\kappa > 0$, given by

$$\overline{F}(x) = \frac{\kappa^\alpha}{(\kappa + x)^\alpha}, \quad x > 0.$$

Another prominent heavy-tailed distribution is the *Weibull distribution* with shape parameter $\tau < 1$ and scale parameter $c > 0$:

$$\overline{F}(x) = e^{-c\,x^\tau}, \quad x > 0.$$

However, for $\tau \geq 1$ the Weibull distribution is light-tailed. We refer to Tables 3.2.17 and 3.2.19 for more distributions used in insurance practice.   □

3.2.3 An Exploratory Statistical Analysis: Mean Excess Plots

The reader might be surprised about the rather arbitrary way in which we discriminated heavy-tailed distributions from light-tailed ones. There are, however, some very good theoretical reasons for the extraordinary role of the exponential distribution as a benchmark distribution, as will be explained in this section.

One tool in order to compare the thickness of the tails of distributions on $[0, \infty)$ is the *mean excess function*.

Definition 3.2.6 (Mean excess function)
Let Y be a non-negative random variable with finite mean, distribution F and $x_l = \inf\{x : F(x) > 0\}$ and $x_r = \sup\{x : F(x) < 1\}$. Then its mean excess *or* mean residual life *function is given by*

$$e_F(u) = E(Y - u \mid Y > u), \quad u \in (x_l, x_r).$$

For our purposes, we mostly consider distributions on $[0, \infty)$ which have support unbounded to the right. The quantity $e_F(u)$ is often referred to as the *mean excess over the threshold value u*. In an insurance context, $e_F(u)$ can be interpreted as the expected claim size in the unlimited layer, over priority u. Here $e_F(u)$ is also called the *mean excess loss* function. In a reliability or medical context, $e_F(u)$ is referred to as the *mean residual life* function. In a financial risk management context, switching from the right tail to the left tail, $e_F(u)$ is referred to as the *expected shortfall*.

The mean excess function of the distribution function F can be written in the form

$$e_F(u) = \frac{1}{\overline{F}(u)} \int_u^\infty \overline{F}(y)\,dy, \quad u > 0. \tag{3.2.10}$$

This formula is often useful for calculations or for deriving theoretical properties of the mean excess function.

Another interesting relationship between e_F and the tail $\overline{F}$ is given by

$$\overline{F}(x) = \frac{e_F(0)}{e_F(x)} \exp\left\{-\int_0^x \frac{1}{e_F(y)}\, dy\right\}, \quad x > 0. \tag{3.2.11}$$

Here we assumed in addition that F is continuous and $F(x) > 0$ for all $x > 0$. Under these additional assumptions, F and e_F determine each other in a unique way. Therefore the tail $\overline{F}$ of a non-negative distribution F and its mean excess function e_F are in a sense equivalent notions. The properties of $\overline{F}$ can be translated into the language of the mean excess function e_F and vice versa.

Derive (3.2.10) and (3.2.11) yourself. Use the relation $EY = \int_0^\infty P(Y > y)\, dy$ which holds for any positive random variable Y.

Example 3.2.7 (Mean excess function of the exponential distribution)
Consider Y with exponential $\mathrm{Exp}(\lambda)$ distribution for some $\lambda > 0$. It is an easy exercise to verify that

$$e_F(u) = \lambda^{-1}, \quad u > 0. \tag{3.2.12}$$

This property is another manifestation of the *forgetfulness* *property* of the exponential distribution; see p. 26. Indeed, the tail of the excess distribution function of Y satisfies

$$P(Y > u + x \mid Y > u) = P(Y > x), \quad x > 0.$$

This means that this distribution function corresponds to an $\mathrm{Exp}(\lambda)$ random variable; it does not depend on the threshold u □

Property (3.2.12) makes the exponential distribution unique: it offers another way of discriminating between heavy- and light-tailed distributions of random variables which are unbounded to the right. Indeed, if $e_F(u)$ converged to infinity for $u \to \infty$, we could call F *heavy-tailed*, if $e_F(u)$ converged to a finite constant as $u \to \infty$, we could call F *light-tailed*. In an insurance context this is quite a sensible definition since unlimited growth of $e_F(u)$ expresses the danger of the underlying distribution F in its right tail, where the large claims come from: given the claim size X_i exceeded the high threshold u, it is very likely that future claim sizes pierce an even higher threshold. On the other hand, for a light-tailed distribution F, the expected excess $E[(X_i - u)_+]$ (here $x_+ = \max(0, x)$) converges to zero (as for the truncated normal distribution) or to a positive constant (as in the exponential case), given $X_i > u$ and the threshold u increases to infinity. This means that claim sizes with light-tailed distributions are much less dangerous (costly) than heavy-tailed distributions.

In Table 3.2.9 we give the mean excess functions of some standard claim size distributions. In Figure 3.2.8 we illustrate the qualitative behavior of $e_F(u)$ for large u.

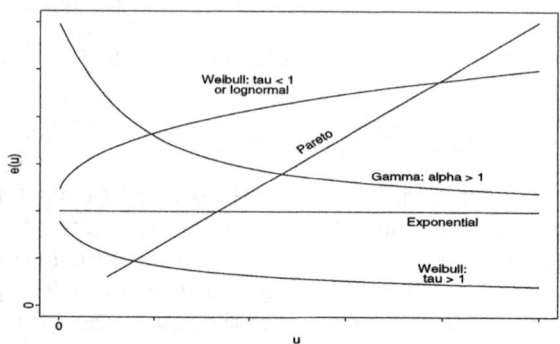

Figure 3.2.8 *Graphs of the mean excess functions $e_F(u)$ for some standard distributions; see Table 3.2.9 for the corresponding parameterizations. Note that heavy-tailed distributions typically have $e_F(u)$ tending to infinity as $u \to \infty$.*

Pareto	$\dfrac{\kappa + u}{\alpha - 1}, \quad \alpha > 1$
Burr	$\dfrac{u}{\alpha\tau - 1}\,(1 + o(1))\,, \quad \alpha\tau > 1$
Log-gamma	$\dfrac{u}{\alpha - 1}\,(1 + o(1))\,, \quad \alpha > 1$
Log-normal	$\dfrac{\sigma^2 u}{\log u - \mu}\,(1 + o(1))$
Benktander type I	$\dfrac{u}{\alpha + 2\beta \log u}$
Benktander type II	$\dfrac{u^{1-\beta}}{\alpha}$
Weibull	$\dfrac{u^{1-\tau}}{c\tau}\,(1 + o(1))$
Exponential	λ^{-1}
Gamma	$\beta^{-1}\left(1 + \dfrac{\alpha - 1}{\beta u} + o\left(\dfrac{1}{u}\right)\right)$
Truncated normal	$u^{-1}\,(1 + o(1))$

Table 3.2.9 *Mean excess functions for some standard distributions. The parameterization is taken from Tables 3.2.17 and 3.2.19. The asymptotic relations are to be understood for $u \to \infty$.*

If one deals with claim size data with an unknown distribution function F, one does not know the mean excess function e_F. As it is often done in statistics, we simply replace F in e_F by its sample version, the empirical distribution function F_n; see (3.2.8). The resulting quantity e_{F_n} is called the *empirical mean excess function*. Since F_n has bounded support, we consider e_{F_n} only for $u \in (X_{(1)}, X_{(n)})$:

$$e_{F_n}(u) = E_{F_n}(Y - u \mid Y > u) = \frac{E_{F_n}(Y - u)_+}{\overline{F}_n(u)}$$

$$= \frac{n^{-1} \sum_{i=1}^{n}(X_i - u)_+}{\overline{F}_n(u)}. \qquad (3.2.13)$$

An alternative expression for e_{F_n} is given by

$$e_{F_n}(u) = \frac{\sum_{i:i \leq n, X_i > u}(X_i - u)}{\#\{i \leq n : X_i > u\}}.$$

An application of the strong law of large numbers to (3.2.13) yields the following result.

Proposition 3.2.10 *Let X_i be iid non-negative random variables with distribution function F which are unbounded to the right. If $EX_1 < \infty$, then for every $u > 0$, $e_{F_n}(u) \stackrel{\text{a.s.}}{\to} e_F(u)$ as $n \to \infty$.*

A graphical test for tail behavior can now be based on e_{F_n}. A *mean excess plot (ME-plot)* consists of the graph

$$\left\{ \left(X_{(k)}, e_{F_n}(X_{(k)}) \right) : k = 1, \ldots, n-1 \right\}.$$

For our purposes, the ME-plot is used *only* as a graphical method, mainly for distinguishing between light- and heavy-tailed models; see Figure 3.2.12 for some simulated examples. Indeed caution is called for when interpreting such plots. Due to the sparseness of the data available for calculating $e_{F_n}(u)$ for large u-values, the resulting plots are very sensitive to changes in the data towards the end of the range; see Figure 3.2.13 for an illustration. For this reason, more robust versions like *median excess plots* and related procedures have been suggested; see for instance Beirlant et al. [10] or Rootzén and Tajvidi [68]. For a critical assessment concerning the use of mean excess functions in insurance, see Rytgaard [69].

Example 3.2.11 (Exploratory data analysis for some real-life data)
In Figures 3.2.5 and 3.2.15 we have graphically summarized some properties of two real-life data sets. The data underlying Figure 3.2.5 correspond to Danish fire insurance claims in millions of Danish Kroner (1985 prices). The data were communicated to us by Mette Rytgaard and correspond to the period 1980-1992, inclusively. There is a total of $n = 2\,493$ observations.

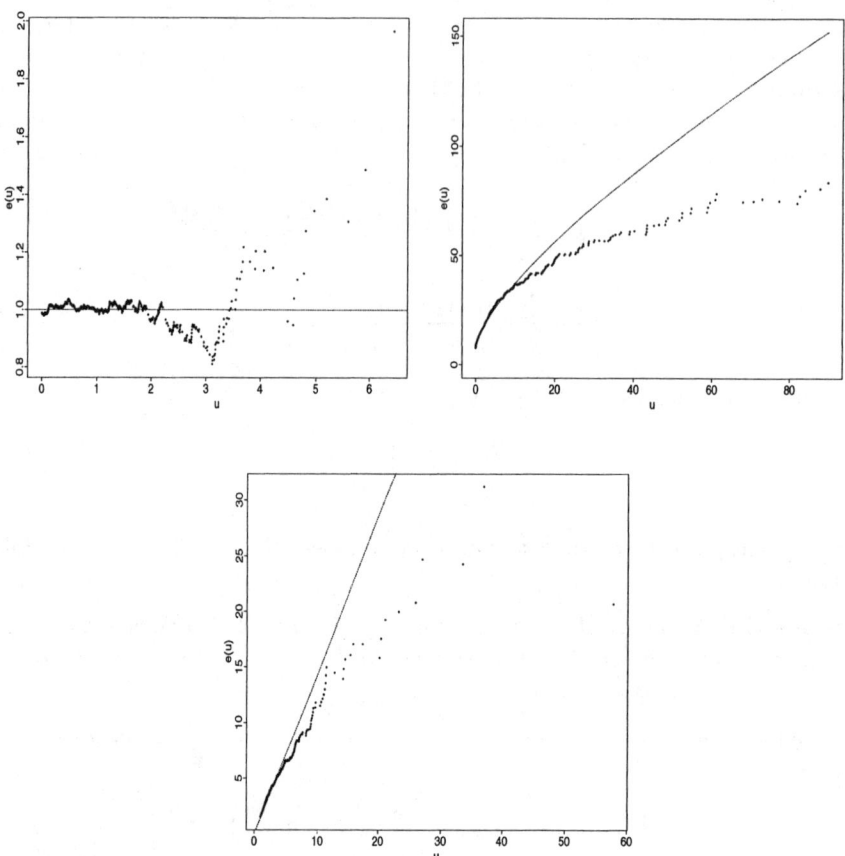

Figure 3.2.12 *The mean excess function plot for* 1 000 *simulated data and the corresponding theoretical mean excess function* e_F *(solid line): standard exponential* (top left), *log-normal* (top right) *with* $\log X \sim N(0,4)$, *Pareto* (bottom) *with tail index* 1.7.

The second insurance data, presented in Figure 3.2.15, correspond to a portfolio of US industrial fire data ($n = 8\,043$) reported over a two year period. This data set is definitely considered by the portfolio manager as "dangerous", i.e., large claim considerations do enter substantially in the final premium calculation.

A first glance at the figures and Table 3.2.14 for both data sets immediately reveals heavy-tailedness and skewedness to the right. The corresponding mean excess functions are close to a straight line which fact indicates that the underlying distributions may be modeled by Pareto-like distribution functions.

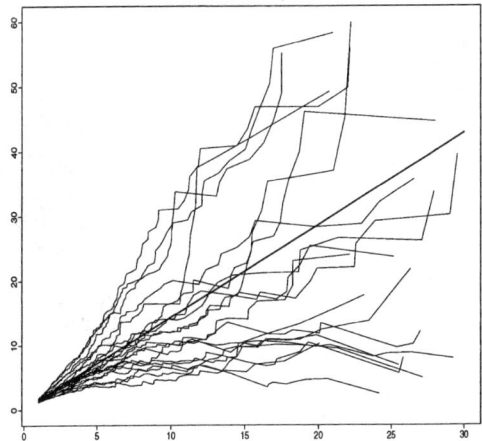

Figure 3.2.13 *The mean excess function of the Pareto distribution* $\overline{F}(x) = x^{-1.7}$, $x \geq 1$, *(straight line), together with* 20 *simulated mean excess plots each based on simulated data* ($n = 1\,000$) *from the above distribution. Note the very unstable behavior, especially towards the higher values of u. This is typical and makes the precise interpretation of* $e_{F_n}(u)$ *difficult; see also Figure 3.2.12.*

The QQ-plots against the standard exponential quantiles also clearly show tails much heavier than exponential ones.

Data	Danish	Industrial
n	2 493	8 043
min	0.313	0.003
1st quartile	1.157	0.587
median	1.634	1.526
mean	3.063	14.65
3rd quartile	2.645	4.488
max	263.3	13 520
$\widehat{x}_{0.99}$	24.61	184.0

Table 3.2.14 *Basic statistics for the Danish and the industrial fire data;* $\widehat{x}_{0.99}$ *stands for the empirical 99%-quantile.*

$\square$

Comments

The importance of the mean excess function (or plot) as a diagnostic tool for insurance data is nicely demonstrated in Hogg and Klugman [44]; see also Beirlant et al. [10] and the references therein.

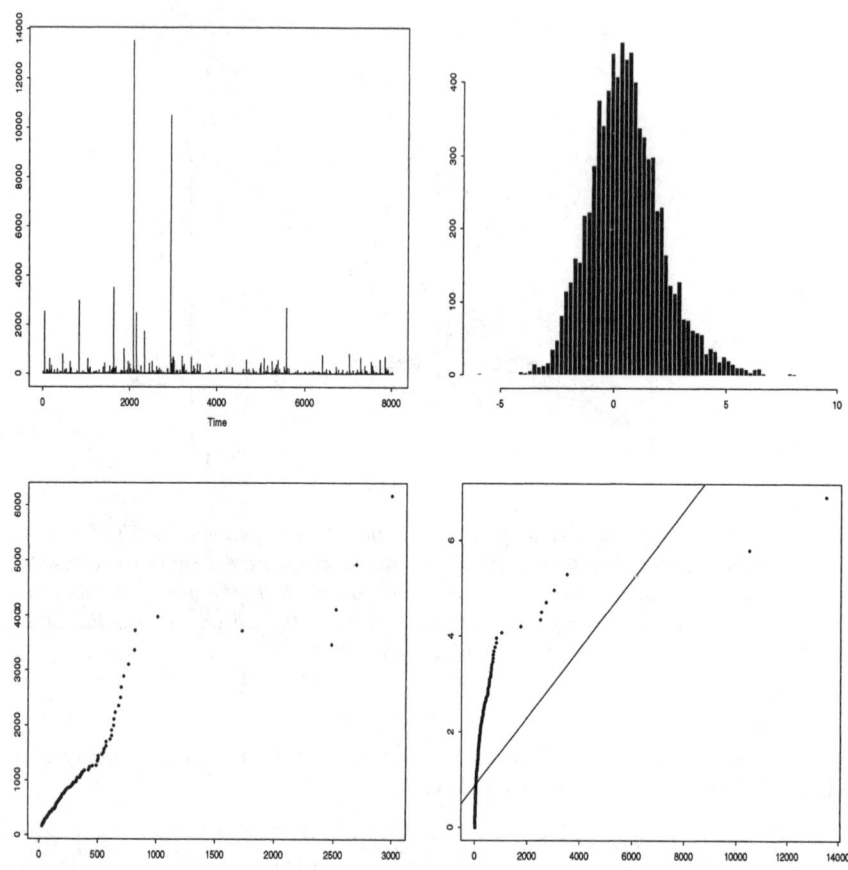

Figure 3.2.15 *Exploratory data analysis of insurance claims caused by industrial fire: the data* (top left), *the histogram of the log-transformed data* (top right), *the ME-plot* (bottom left) *and a QQ-plot against standard exponential quantiles* (bottom right). *See Example 3.2.11 for some comments.*

3.2.4 Standard Claim Size Distributions and Their Properties

Classical non-life insurance mathematics was most often concerned with claim size distributions with light tails in the sense which has been made precise in Section 3.2.3. We refer to Table 3.2.17 for a collection of such distributions. These distributions have mean excess functions $e_F(u)$ converging to some finite limit as $u \to \infty$, provided the support is infinite. For obvious reasons, we call them *small claim distributions*. One of the main reasons for the popularity of these distributions is that they are standard distributions in statistics. Classical statistics deals with the normal and the gamma distributions,

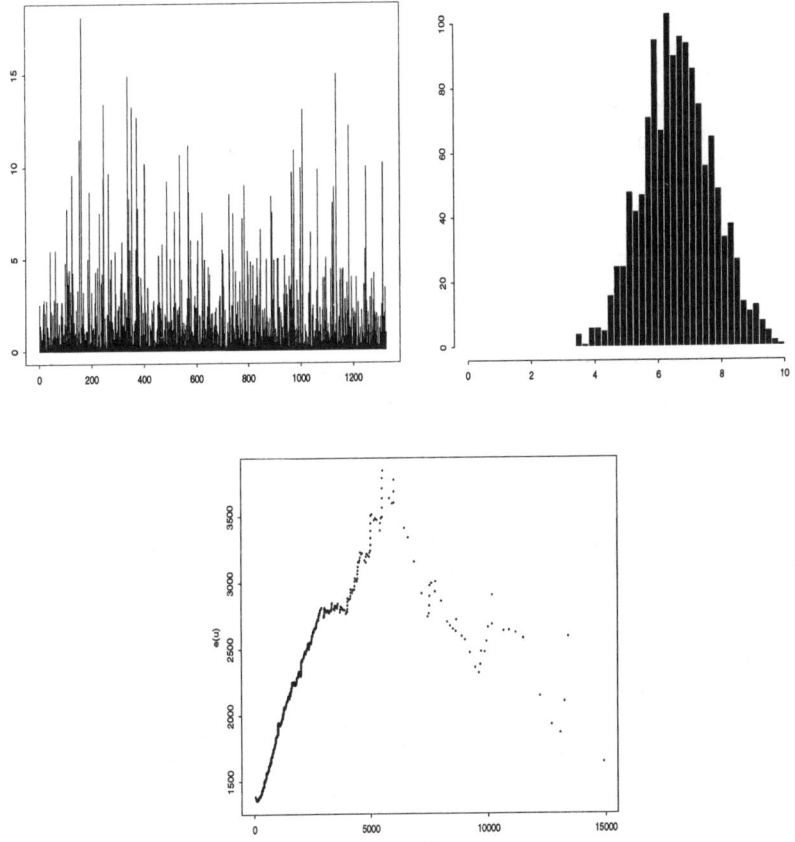

Figure 3.2.16 *Exploratory data analysis of insurance claims caused by water: the data* (top, left), *the histogram of the log-transformed data* (top, right), *the ME-plot* (bottom). *Notice the kink in the ME-plot in the range* (5 000, 6 000) *reflecting the fact that the data seem to cluster towards some specific upper value.*

among others, and in any introductory course on statistics we learn about these distributions because they have certain optimality conditions (closure of the normal and gamma distributions under convolutions, membership in exponential families, etc.) and therefore we can apply standard estimation techniques such as maximum likelihood.

In Figure 3.2.16 one can find a claim size sample which one could model by one of the distributions from Table 3.2.17. Indeed, notice that the mean excess plot of these data curves down at the right end, indicating that the right tail of the underlying distribution is not too dangerous. It is also common practice to fit distributions with bounded support to insurance claim data, for example by

Name	Tail $\overline{F}$ or density f	Parameters
Exponential	$\overline{F}(x) = \mathrm{e}^{-\lambda x}$	$\lambda > 0$
Gamma	$f(x) = \dfrac{\beta^\alpha}{\Gamma(\alpha)}\, x^{\alpha-1}\mathrm{e}^{-\beta x}$	$\alpha, \beta > 0$
Weibull	$\overline{F}(x) = \mathrm{e}^{-cx^\tau}$	$c > 0,\, \tau \geq 1$
Truncated normal	$f(x) = \sqrt{\dfrac{2}{\pi}}\, \mathrm{e}^{-x^2/2}$	—
Any distribution with bounded support		

Table 3.2.17 *Claim size distributions : "small claims".*

truncating any of the heavy-tailed distributions in Table 3.2.19 at a certain upper limit. This makes sense if the insurer has to cover claim sizes only up to this upper limit or for a certain layer. In this situation it is, however, reasonable to use the full data set (not just the truncated data) for estimating the parameters of the distribution.

Over the last few years the (re-)insurance industry has faced new challenges due to climate change, pollution, riots, earthquakes, terrorism, etc. We refer to Table 3.2.18 for a collection of the largest insured losses 1970-2002, taken from *Sigma* [73]. For this kind of data one would not use the distributions of Table 3.2.17, but rather those presented in Table 3.2.19. All distributions of this table are heavy-tailed in the sense that their mean excess functions $e_F(u)$ increase to infinity as $u \to \infty$; cf. Table 3.2.9. As a matter of fact, the distributions of Table 3.2.19 are not easily fitted since various of their characteristics (such as the tail index α of the Pareto distribution) can be estimated only by using the largest upper order statistics in the sample. In this case, *extreme value statistics* is called for. This means that, based on theoretical (semi-)parametric models from extreme value theory such as the extreme value distributions and the generalized Pareto distribution, one needs to fit those distributions from a relatively small number of upper order statistics or from the excesses of the underlying data over high thresholds. We refer to Embrechts et al. [29] for an introduction to the world of extremes.

We continue with some more specific comments on the distributions in Table 3.2.19. Perhaps with the exception of the log-normal distribution, these distributions are not most familiar from a standard course on statistics or probability theory.

The Pareto, Burr, log-gamma and truncated α-stable distributions have in common that their right tail is of the asymptotic form

Losses	Date	Event	Country
20 511	08/24/92	Hurricane "Andrew"	US, Bahamas
19 301	09/11/01	Terrorist attack on WTC, Pentagon and other buildings	US
16 989	01/17/94	Northridge earthquake in California	US
7 456	09/27/91	Tornado "Mireille"	Japan
6 321	01/25/90	Winter storm "Daria"	Europe
6 263	12/25/99	Winter storm "Lothar"	Europe
6 087	09/15/89	Hurricane "Hugo"	P. Rico, US
4 749	10/15/87	Storm and floods	Europe
4 393	02/26/90	Winter storm "Vivian"	Europe
4 362	09/22/99	Typhoon "Bart" hits the south of the country	Japan
3 895	09/20/98	Hurricane "Georges"	US, Caribbean
3 200	06/05/01	Tropical storm "Allison"; flooding	US
3 042	07/06/88	Explosion on "Piper Alpha" offshore oil rig	UK
2 918	01/17/95	Great "Hanshin" earthquake in Kobe	Japan
2 592	12/27/99	Winter storm "Martin"	France, Spain, CH
2 548	09/10/99	Hurricane "Floyd", heavy down-pours, flooding	US, Bahamas
2 500	08/06/02	Rains, flooding	Europe
2 479	10/01/95	Hurricane "Opal"	US, Mexico
2 179	03/10/93	Blizzard, tornadoes	US, Mexico, Canada
2 051	09/11/92	Hurricane "Iniki"	US, North Pacific
1 930	04/06/01	Hail, floods and tornadoes	US
1 923	10/23/89	Explosion at Philips Petroleum	US
1 864	09/03/79	Hurricane "Frederic"	US
1 835	09/05/96	Hurricane "Fran"	US
1 824	09/18/74	Tropical cyclone "Fifi"	Honduras
1 771	09/03/95	Hurricane "Luis"	Caribbean
1 675	04/27/02	Spring storm with several tornadoes	US
1 662	09/12/88	Hurricane "Gilbert"	Jamaica
1 620	12/03/99	Winter storm "Anatol"	Europe
1 604	05/03/99	Series of 70 tornadoes in the Midwest	US
1 589	12/17/83	Blizzard, cold wave	US, Mexico, Canada
1 585	10/20/91	Forest fire which spread to urban area	US
1 570	04/02/74	Tornados in 14 states	US
1 499	04/25/73	Flooding on the Mississippi	US
1 484	05/15/98	Wind, hail and tornadoes (MN, IA)	US
1 451	10/17/89	"Loma Prieta" earthquake	US
1 436	08/04/70	Hurricane "Celia"	US
1 409	09/19/98	Typhoon "Vicki"	Japan, Philippines
1 358	01/05/98	Cold spell with ice and snow	Canada, US
1 340	05/05/95	Wind, hail and flooding	US

Table 3.2.18 *The* 40 *most costly insurance losses* 1970 − 2002. *Losses are in million* $US *indexed to* 2002 *prices. The table is taken from* Sigma [73] *with friendly permission of Swiss Re Zurich.*

Name	Tail $\overline{F}$ or density f	Parameters
Log-normal	$f(x) = \dfrac{1}{\sqrt{2\pi}\,\sigma x}\mathrm{e}^{-(\log x - \mu)^2/(2\sigma^2)}$	$\mu \in \mathbb{R},\ \sigma > 0$
Pareto	$\overline{F}(x) = \left(\dfrac{\kappa}{\kappa + x}\right)^{\alpha}$	$\alpha, \kappa > 0$
Burr	$\overline{F}(x) = \left(\dfrac{\kappa}{\kappa + x^{\tau}}\right)^{\alpha}$	$\alpha, \kappa, \tau > 0$
Benktander type I	$\overline{F}(x) = (1 + 2(\beta/\alpha)\log x)$ $\mathrm{e}^{-\beta(\log x)^2 - (\alpha+1)\log x}$	$\alpha, \beta > 0$
Benktander type II	$\overline{F}(x) = \mathrm{e}^{\alpha/\beta} x^{-(1-\beta)} \mathrm{e}^{-\alpha\, x^{\beta}/\beta}$	$\alpha > 0$ $0 < \beta < 1$
Weibull	$\overline{F}(x) = \mathrm{e}^{-cx^{\tau}}$	$c > 0$ $0 < \tau < 1$
Log-gamma	$f(x) = \dfrac{\alpha^{\beta}}{\Gamma(\beta)}(\log x)^{\beta-1}\, x^{-\alpha-1}$	$\alpha, \beta > 0$
Truncated α-stable	$\overline{F}(x) = P(|X| > x)$ where X is an α-stable random variable	$1 < \alpha < 2$

Table 3.2.19 *Claim size distributions : "large claims". All distributions have support $(0, \infty)$ except for the Benktander cases and the log-gamma with $(1, \infty)$. For the definition of an α-stable distribution, see Embrechts et al. [29], p. 71; cf. Exercise 16 on p. 56.*

$$\lim_{x \to \infty} \frac{\overline{F}(x)}{x^{-\alpha}(\log x)^{\gamma}} = c,$$

for some constants $\alpha, c > 0$ and $\gamma \in \mathbb{R}$. Tails of this kind are called *regularly varying*. We will come back to this notion in Section 3.2.5.

The log-gamma, Pareto and log-normal distributions are obtained by an exponential transformation of a random variable with gamma, exponential and normal distribution, respectively. For example, let Y be $N(\mu, \sigma^2)$ distributed. Then $\exp\{Y\}$ has the log-normal distribution with density given in Table 3.2.19. The goal of these exponential transformations of random variables with a standard *light-tailed* distribution is to create *heavy-tailed* distributions in a simple way. An advantage of this procedure is that by a logarithmic transformation of the data one returns to the standard light-tailed distributions. In particular, one can use standard theory for the estimation of the underlying parameters.

Some of the distributions in Table 3.2.19 were introduced as extensions of the Pareto, log-normal and Weibull ($\tau < 1$) distributions as classical heavy-tailed distributions. For example, the Burr distribution differs from the Pareto distribution only by the additional shape parameter τ. As a matter of fact, practice in extreme value statistics (see for example Chapter 6 in Embrechts et al. [29], or convince yourself by a simulation study) shows that it is hard, if not impossible, to distinguish between the log-gamma, Pareto, Burr distributions based on parameter (for example maximum likelihood) estimation. It is indeed difficult to estimate the tail parameter α, the shape parameter τ or the scale parameter κ accurately in any of the cases. Similar remarks apply to the Benktander type I and the log-normal distributions, as well as the Benktander type II and the Weibull ($\tau < 1$) distributions. The Benktander distributions were introduced in the insurance world for one particular reason: one can explicitly calculate their mean excess functions; cf. Table 3.2.9.

3.2.5 Regularly Varying Claim Sizes and Their Aggregation

Although the distribution functions F in Table 3.2.19 look different, some of them are quite similar with regard to their asymptotic tail behavior. Those include the Pareto, Burr, stable and log-gamma distributions. In particular, their right tails can be written in the form

$$\overline{F}(x) = 1 - F(x) = \frac{L(x)}{x^\alpha}, \quad x > 0,$$

for some constant $\alpha > 0$ and a positive measurable function $L(x)$ on $(0, \infty)$ satisfying

$$\lim_{x \to \infty} \frac{L(cx)}{L(x)} = 1 \quad \text{for all } c > 0. \tag{3.2.14}$$

A function with this property is called *slowly varying* (at infinity). Examples of such functions are:

constants, logarithms, powers of logarithms, iterated logarithms.

Every slowly varying function has the representation

$$L(x) = c_0(x) \exp\left\{ \int_{x_0}^x \frac{\varepsilon(t)}{t} \, dt \right\}, \quad \text{for } x \geq x_0, \text{ some } x_0 > 0, \tag{3.2.15}$$

where $\varepsilon(t) \to 0$ as $t \to \infty$ and $c_0(t)$ is a positive function satisfying $c_0(t) \to c_0$ for some positive constant c_0. Using representation (3.2.15), one can show that for every $\delta > 0$,

$$\lim_{x \to \infty} \frac{L(x)}{x^\delta} = 0 \quad \text{and} \quad \lim_{x \to \infty} x^\delta L(x) = \infty, \tag{3.2.16}$$

i.e., L is "small" compared to any power function, x^δ.

Definition 3.2.20 (Regularly varying function and regularly varying random variable)

Let L be a slowly varying function in the sense of (3.2.14) .

(1) *For any $\delta \in \mathbb{R}$, the function*

$$f(x) = x^\delta L(x), \quad x > 0,$$

is said to be regularly varying with index δ.

(2) *A positive random variable X and its distribution are said to be* regularly varying[2] *with (tail) index $\alpha \geq 0$ if the right tail of the distribution has the representation*

$$P(X > x) = L(x)\, x^{-\alpha}, \quad x > 0.$$

An alternative way of defining regular variation with index δ is to require

$$\lim_{x \to \infty} \frac{f(cx)}{f(x)} = c^\delta \quad \text{for all } c > 0. \tag{3.2.17}$$

Regular variation is one possible way of describing "small" deviations from exact power law behavior. It is hard to believe that social or natural phenomena can be described by exact power law behavior. It is, however, known that various phenomena, such as Zipf's law, fractal dimensions, the probability of exceedances of high thresholds by certain iid data, the world income distribution, etc., can be well described by functions which are "almost power" functions; see Schroeder [72] for an entertaining study of power functions and their application to different scaling phenomena. Regular variation is an appropriate concept in this context. It has been carefully studied for many years and arises in different areas, such as summation theory of independent or weakly dependent random variables, or in extreme value theory as a natural condition on the tails of the underlying distributions. We refer to Bingham et al. [14] for an encyclopedic treatment of regular variation.

Regularly varying distributions with positive index, such as the Pareto, Burr, α-stable, log-gamma distributions, are claim size distributions with some of the heaviest tails which have ever been fitted to claim size data. Although it is theoretically possible to construct distributions with tails which are heavier than any power law, statistical evidence shows that there is no need for such distributions. As as a matter of fact, if X is regularly varying with index $\alpha > 0$, then

$$EX^\delta \begin{cases} = \infty & \text{for } \delta > \alpha, \\ < \infty & \text{for } \delta < \alpha, \end{cases}$$

---

[2] This definition differs from the standard usage of the literature which refers to X as a random variable with regularly varying tail and to its distribution as distribution with regularly varying tail.

i.e., moments below order α are finite, and moments above α are infinite.[3] (Verify these moment relations by using representation (3.2.15).) The value α can be rather low for claim sizes occurring in the context of reinsurance. It is not atypical that α is below 2, sometimes even below 1, i.e., the variance or even the expectation of the distribution fitted to the data can be infinite. We refer to Example 3.2.11 for two data sets, where statistical estimation procedures provide evidence for values α close to or even below 2; see Chapter 6 in Embrechts et al. [29] for details.

As we have learned in the previous sections, one of the important quantities in insurance mathematics is the total claim amount $S(t) = \sum_{i=1}^{N(t)} X_i$. It is a random partial sum process with iid positive claim sizes X_i as summands, independent of the claim number process N. A complicated but important practical question is to get exact formulae or good approximations (by numerical or Monte Carlo methods) to the distribution of $S(t)$. Later in this course we will touch upon this problem; see Section 3.3.

In this section we focus on a simpler problem: the tail asymptotics of the distribution of the first n aggregated claim sizes

$$S_n = X_1 + \cdots + X_n \,, \quad n \geq 1 \,.$$

We want to study how heavy tails of the claim size distribution function F influence the tails of the distribution function of S_n. From a reasonable notion of heavy-tailed distributions we would expect that the heavy tails do not disappear by aggregating independent claim sizes. This is exactly the content of the following result.

Lemma 3.2.21 *Assume that X_1 and X_2 are independent regularly varying random variables with the same index $\alpha > 0$, i.e.,*

$$\overline{F}_i(x) = P(X_i > x) = \frac{L_i(x)}{x^\alpha}\,, \quad x > 0\,.$$

for possibly different slowly varying functions L_i. Then $X_1 + X_2$ is regularly varying with the same index. More precisely, as $x \to \infty$,

$$P(X_1 + X_2 > x) = [P(X_1 > x) + P(X_2 > x)]\,(1 + o(1))$$

$$= x^{-\alpha}\,[L_1(x) + L_2(x)]\,(1 + o(1))\,.$$

Proof. Write $G(x) = P(X_1 + X_2 \leq x)$ for the distribution function of $X_1 + X_2$. Using $\{X_1 + X_2 > x\} \supset \{X_1 > x\} \cup \{X_2 > x\}$, one easily checks that

$$\overline{G}(x) \geq \left(\overline{F}_1(x) + \overline{F}_2(x)\right)(1 - o(1))\,.$$

---

[3] These moment relations do not characterize a regularly varying distribution. A counterexample is the Peter-and-Paul distribution with distribution function $F(x) = \sum_{k \geq 1 : 2^k \leq x} 2^{-k}$, $x \geq 0$. This distribution has finite moments of order $\delta < 1$ and infinite moments of order $\delta \geq 1$, but it is not regularly varying with index 1. See Exercise 7 on p. 114.

If $0 < \delta < 1/2$, then from

$$\{X_1 + X_2 > x\} \subset \{X_1 > (1 - \delta)x\} \cup \{X_2 > (1 - \delta)x\} \cup \{X_1 > \delta x, X_2 > \delta x\} ,$$

it follows that

$$\overline{G}(x) \le \overline{F}_1((1 - \delta)x) + \overline{F}_2((1 - \delta)x) + \overline{F}_1(\delta x)\,\overline{F}_2(\delta x)$$

$$= \Big(\overline{F}_1((1 - \delta)x) + \overline{F}_2((1 - \delta)x)\Big)(1 + o(1)) .$$

Hence

$$1 \le \liminf_{x \to \infty} \frac{\overline{G}(x)}{\overline{F}_1(x) + \overline{F}_2(x)} \le \limsup_{x \to \infty} \frac{\overline{G}(x)}{\overline{F}_1(x) + \overline{F}_2(x)} \le (1 - \delta)^{-\alpha} ,$$

and the result is established upon letting $\delta \downarrow 0$. □

An important corollary, obtained via induction on n, is the following:

Corollary 3.2.22 *Assume that $X_1, \ldots, X_n$ are n iid regularly varying random variables with index $\alpha > 0$ and distribution function F. Then S_n is regularly varying with index n, and*

$$P(S_n > x) = n\,\overline{F}(x)\,(1 + o(1)), \qquad x \to \infty .$$

Suppose now that $X_1, \ldots, X_n$ are iid with distribution function F, as in the above corollary. Denote the partial sum of $X_1, \ldots, X_n$ by $S_n = X_1 + \cdots + X_n$ and their partial maximum by $M_n = \max(X_1, \ldots, X_n)$. Then for $n \ge 2$ as $x \to \infty$,

$$P(M_n > x) = \overline{F^n}(x) = \overline{F}(x) \sum_{k=0}^{n-1} F^k(x) = n\,\overline{F}(x)\,(1 + o(1)) .$$

Therefore, with the above notation, Corollary 3.2.22 can be reformulated as: if X_i is regularly varying with index $\alpha > 0$ then

$$\lim_{x \to \infty} \frac{P(S_n > x)}{P(M_n > x)} = 1, \quad \text{for } n \ge 2.$$

This implies that for distributions with regularly varying tails, the tail of the distribution of the sum S_n is essentially determined by the tail of the distribution of the maximum M_n. This is in fact one of the intuitive notions of heavy-tailed or large claim distributions. Hence, stated in a somewhat vague way: *under the assumption of regular variation, the tail of the distribution of the maximum claim size determines the tail of the distribution of the aggregated claim sizes.*

Comments

Surveys on regularly varying functions and distributions can be found in many standard textbooks on probability theory and extreme value theory; see for example Feller [32], Embrechts et al. [29] or Resnick [64]. The classical reference to regular variation is the book by Bingham et al. [14].

3.2.6 Subexponential Distributions

We learned in the previous section that for iid regularly varying random variables $X_1, X_2, \ldots$ with positive index α, the tail of the sum $S_n = X_1 + \cdots + X_n$ is essentially determined by the tail of the maximum $M_n = \max_{i=1,\ldots,n} X_i$. To be precise, we found that $P(S_n > x) = P(M_n > x)(1 + o(1))$ as $x \to \infty$ for every $n = 1, 2, \ldots$. The latter relation can be taken as a natural definition for "heavy-tailedness" of a distribution:

Definition 3.2.23 (Subexponential distribution)
The positive random variable X and its distribution are said to be subexponential if for a sequence (X_i) of iid random variables with the same distribution as X the following relation holds:

$$\text{For all } n \geq 2: \quad P(S_n > x) = P(M_n > x)(1 + o(1)), \quad \text{as } x \to \infty.$$
$$(3.2.18)$$

The class of subexponential distributions is denoted by $\mathcal{S}$.

One can show that the defining property (3.2.18) holds for *all* $n \geq 2$ if it holds for *some* $n \geq 2$; see Section 1.3.2 in [29] for details.

As we have learned in Section 3.2.5, $P(M_n > x) = n\overline{F}(x)(1 + o(1))$ as $x \to \infty$, where F is the common distribution function of the X_i's, and therefore the defining property (3.2.18) can also be formulated as

$$\text{For all } n \geq 2: \quad \lim_{x \to \infty} \frac{P(S_n > x)}{\overline{F}(x)} = n.$$

We consider some properties of subexponential distributions.

Lemma 3.2.24 (Basic properties of subexponential distributions)

(1) *If $F \in \mathcal{S}$, then for any $y > 0$,*

$$\lim_{x \to \infty} \frac{\overline{F}(x - y)}{\overline{F}(x)} = 1.$$
$$(3.2.19)$$

(2) *If (3.2.19) holds then, for all $\varepsilon > 0$,*

$$e^{\varepsilon x} \overline{F}(x) \to \infty, \quad x \to \infty.$$

(3) *If $F \in S$ then, given $\varepsilon > 0$, there exists a finite constant K so that for all $n \geq 2$,*

$$\frac{P(S_n > x)}{\overline{F}(x)} \leq K \,(1+\varepsilon)^n \,, \qquad x \geq 0 \,. \tag{3.2.20}$$

For the proof of (3), see Lemma 1.3.5 in [29].

Proof. (1) Write $G(x) = P(X_1 + X_2 \leq x)$ for the distribution function of $X_1 + X_2$. For $x \geq y > 0$,

$$\frac{\overline{G}(x)}{\overline{F}(x)} = 1 + \int_0^y \frac{\overline{F}(x-t)}{\overline{F}(x)}\, dF(t) + \int_y^x \frac{\overline{F}(x-t)}{\overline{F}(x)}\, dF(t)$$

$$\geq 1 + F(y) + \frac{\overline{F}(x-y)}{\overline{F}(x)}\, (F(x) - F(y)) \,.$$

Thus, if x is large enough so that $F(x) - F(y) \neq 0$,

$$1 \leq \frac{\overline{F}(x-y)}{\overline{F}(x)} \leq \left(\frac{\overline{G}(x)}{\overline{F}(x)} - 1 - F(y)\right) (F(x) - F(y))^{-1} \,.$$

In the latter estimate, the right-hand side tends to 1 as $x \to \infty$. This proves (3.2.19).

(2) By virtue of (1), the function $\overline{F}(\log y)$ is slowly varying. But then the conclusion that $y^\varepsilon \overline{F}(\log y) \to \infty$ as $y \to \infty$ follows immediately from the representation theorem for slowly varying functions; see (3.2.16). Now write $y = e^x$. □

Lemma 3.2.24(2) justifies the name "subexponential" for $F \in S$; indeed $\overline{F}(x)$ decays to 0 slower than any exponential function $e^{-\varepsilon x}$ for $\varepsilon > 0$. Furthermore, since for any $\varepsilon > 0$,

$$E e^{\varepsilon X} \geq E(e^{\varepsilon X}\, I_{(y,\infty)}) \geq e^{\varepsilon y}\, \overline{F}(y) \,, \qquad y \geq 0 \,,$$

it follows from Lemma 3.2.24(2) that for $F \in S$, $E e^{\varepsilon X} = \infty$ for all $\varepsilon > 0$. Therefore the moment generating function of a subexponential distribution does not exist in any neighborhood of the origin.

   Property (3.2.19) holds for larger classes of distributions than the subexponential distributions. It can be taken as another definition of heavy-tailed distributions. It means that the tails $P(X > x)$ and $P(X + y > x)$ are not significantly different, for any fixed y and large x. In particular, it says that for any $y > 0$ as $x \to \infty$,

$$\frac{P(X > x+y)}{P(X > x)} = \frac{P(X > x+y, X > x)}{P(X > x)}$$

$$= P(X > x+y \mid X > x) \to 1 \,. \tag{3.2.21}$$

Thus, once X has exceeded a high threshold, x, it is very likely to exceed an even higher threshold $x + y$. This situation changes completely when we look, for example, at an exponential or a truncated normal random variable. For these two distributions you can verify that the above limit exists, but its value is less than 1.

Property (3.2.19) helps one to exclude certain distributions from the class $\mathcal{S}$. However, it is in general difficult to determine whether a given distribution is subexponential.

Example 3.2.25 (Examples of subexponential distributions)
The large claim distributions in Table 3.2.19 are subexponential. The small claim distributions in Table 3.2.17 are not subexponential. However, the tail of a subexponential distribution can be very close to an exponential distribution. For example, the heavy-tailed Weibull distributions with tail

$$\overline{F}(x) = e^{-c\,x^{\tau}}, \quad x \geq 0, \quad \text{for some } \tau \in (0,1),$$

and also the distributions with tail

$$\overline{F}(x) = e^{-x\,(\log x)^{-\beta}}, \quad x \geq x_0, \quad \text{for some } \beta, x_0 > 0,$$

are subexponential. We refer to Sections 1.4.1 and A3.2 in [29] for details. See also Exercise 11 on p. 114.   □

Comments

The subexponential distributions constitute a natural class of heavy-tailed claim size distributions from a theoretical but also from a practical point of view. In insurance mathematics subexponentiality is considered as a synonym for heavy-tailedness. The class $\mathcal{S}$ is very flexible insofar that it contains distributions with very heavy tails such as the regularly varying subclass, but also distributions with moderately heavy tails such as the log-normal and Weibull ($\tau < 1$) distributions. In contrast to regularly varying random variables, log-normal and Weibull distributed random variables have finite power moments, but none of the subexponential distributions has a finite moment generating function in some neighborhood of the origin.

An extensive treatment of subexponential distributions, their properties and use in insurance mathematics can be found in Embrechts et al. [29]. A more recent survey on $\mathcal{S}$ and related classes of distributions is given in Goldie and Klüppelberg [35].

We re-consider subexponential claim size distributions when we study ruin probabilities in Section 4.2.4. There subexponential distributions will turn out to be the most natural class of large claim distributions.

Exercises

Section 3.2.2

(1) We say that a distribution is light-tailed (compared to the exponential distribution) if

$$\limsup_{x \to \infty} \frac{\overline{F}(x)}{e^{-\lambda x}} < \infty$$

for some $\lambda > 0$ and heavy-tailed if

$$\liminf_{x \to \infty} \frac{\overline{F}(x)}{e^{-\lambda x}} > 0$$

for all $\lambda > 0$.

(a) Show that the gamma and the truncated normal distributions are light-tailed.

(b) Consider a Pareto distribution given via its tail in the parameterization

$$\overline{F}(x) = \frac{\kappa^\alpha}{(\kappa + x)^\alpha}, \quad x > 0. \tag{3.2.22}$$

Show that F is heavy-tailed.

(c) Show that the Weibull distribution with tail $\overline{F}(x) = e^{-cx^\tau}$, $x > 0$, for some $c, \tau > 0$, is heavy-tailed for $\tau < 1$ and light-tailed for $\tau \geq 1$.

Section 3.2.3

(2) Let F be the distribution function of a positive random variable X with infinite right endpoint, finite expectation and $F(x) > 0$ for all $x > 0$.

(a) Show that the mean excess function e_F satisfies the relation

$$e_F(x) = \frac{1}{\overline{F}(x)} \int_x^\infty \overline{F}(y)\,dy, \quad x > 0.$$

(b) A typical heavy-tailed distribution is the Pareto distribution given via its tail in the parameterization

$$\overline{F}(x) = \gamma^\alpha x^{-\alpha}, \quad x > \gamma, \tag{3.2.23}$$

for positive γ and α. Calculate the mean excess function e_F for $\alpha > 1$ and verify that $e_F(x) \to \infty$ as $x \to \infty$. Why do we need the condition $\alpha > 1$?

(c) Assume F is continuous and has support $(0, \infty)$. Show that

$$\overline{F}(x) = \frac{e_F(0)}{e_F(x)} \exp\left\{-\int_0^x (e_F(y))^{-1}\,dy\right\}, \quad x > 0.$$

Hint: Interpret $-1/e_F(y)$ as logarithmic derivative.

(3) The *generalized Pareto distribution* plays a major role in extreme value theory and extreme value statistics; see Embrechts et al. [29], Sections 3.4 and 6.5. It is given by its distribution function

$$G_{\xi,\beta}(x) = 1 - \left(1 + \xi \frac{x}{\beta}\right)^{-1/\xi}, \quad x \in D(\xi, \beta).$$

Here $\xi \in \mathbb{R}$ is a shape parameter and $\beta > 0$ a scale parameter. For $\xi = 0$, $G_{0,\beta}(x)$ is interpreted as the limiting distribution as $\xi \to 0$:

$$G_{0,\beta}(x) = 1 - e^{-x/\beta}.$$

The domain $D(\xi, \beta)$ is defined as follows:

$$D(\xi, \beta) = \begin{cases} [0, \infty) & \xi \geq 0, \\ [0, -1/\xi] & \xi < 0. \end{cases}$$

Show that $G_{\xi,\beta}$ has the mean excess function

$$e_G(u) = \frac{\beta + \xi u}{1 - \xi}, \qquad \beta + u\xi > 0,$$

for u in the support of $G_{\xi,\beta}$ and $\xi < 1$.

Sections 3.2.4-3.2.5

(4) Some properties of Pareto-like distributions.

(a) Verify for a random variable X with Pareto distribution function F given by (3.2.22) that $EX^\delta = \infty$ for $\delta \geq \alpha$ and $EX^\alpha < \infty$ for $\delta < \alpha$.

(b) Show that a Pareto distributed random variable X whose distribution has parameterization (3.2.23) is obtained by the transformation $X = (\gamma \exp\{Y\})^{1/\alpha}$ for some standard exponential random variable Y and $\gamma, \alpha > 0$.

(c) A Burr distributed random variable Y is obtained by the transformation $Y = X^{1/c}$ for some positive c from a Pareto distributed random variable X with tail (3.2.22). Determine the tail $\overline{F}_Y(x)$ for the Burr distribution and check for which $p > 0$ the moment EY^p is finite.

(d) The log-gamma distribution has density

$$f(y) = \frac{\delta^\gamma \lambda^\delta}{\Gamma(\gamma)} \frac{(\log(y/\lambda))^{\gamma-1}}{y^{\delta+1}}, \qquad y > \lambda.$$

for some $\lambda, \gamma, \delta > 0$. Check by some appropriate bounds for $\log x$ that the log-gamma distribution has finite moments of order less than δ and infinite moments of order greater than δ. Check that the tail $\overline{F}$ satisfies

$$\lim_{x \to \infty} \frac{\overline{F}(x)}{(\delta^{\gamma-1}\lambda^\delta/\Gamma(\gamma))(\log(x/\lambda))^{\gamma-1}x^{-\delta}} = 1.$$

(e) Let X have a Pareto distribution with tail (3.2.23). Consider a positive random variable $Y > 0$ with $EY^\alpha < \infty$. Show that

$$\lim_{x \to \infty} \frac{P(XY > x)}{P(X > x)} = EY^\alpha.$$

Hint: Use a conditioning argument.

(5) Consider the Pareto distribution in the parameterization (3.2.23), where we assume the constant γ to be known. Determine the maximum likelihood estimator of α based on an iid sample $X_1, \ldots, X_n$ with distribution function F and the distribution of $1/\hat{\alpha}_{\text{MLE}}$. Why is this result not surprising? See (4,b).

(6) Recall the representation (3.2.15) of a slowly varying function.

(a) Show that (3.2.15) defines a slowly varying function.

(b) Use representation (3.2.15) to show that for any slowly varying function L and $\delta > 0$, the properties $\lim_{x \to \infty} x^\delta L(x) = \infty$ and $\lim_{x \to \infty} x^{-\delta} L(x) = 0$ hold.

(7) Consider the Peter-and-Paul distribution function given by

$$F(x) = \sum_{k \geq 1:\, 2^k \leq x} 2^{-k}, \quad x \geq 0.$$

(a) Show that $\overline{F}$ is not regularly varying.
(b) Show that for a random variable X with distribution function F, $EX^\delta = \infty$ for $\delta \geq 1$ and $EX^\delta < \infty$ for $\delta < 1$.

Section 3.2.6

(8) Show by different means that the exponential distribution is not subexponential.
(a) Verify that the defining property (3.2.18) of a subexponential distribution does not hold.
(b) Verify that condition (3.2.21) does not hold. The latter condition is necessary for subexponentiality.
(c) Use an argument about the exponential moments of a subexponential distribution.
(9) Show that the light-tailed Weibull distribution given by $\overline{F}(x) = e^{-c x^\tau}$, $x > 0$, for some $c > 0$ and $\tau \geq 1$ is not subexponential.
(10) Show that a claim size distribution with finite support cannot be subexponential.
(11) Pitman [62] gave a complete characterization of subexponential distribution functions F with a density f in terms of their *hazard rate function* $q(x) = f(x)/\overline{F}(x)$. In particular, he showed the following.

Assume that $q(x)$ is eventually decreasing to 0. Then
(i) $F \in \mathcal{S}$ if and only if

$$\lim_{x \to \infty} \int_0^x e^{y\, q(y)} f(y)\, dy = 1.$$

(ii) If the function $g(x) = e^{x\, q(x)} f(x)$ is integrable on $[0, \infty)$, then $F \in \mathcal{S}$.

Apply these results in order to show that the distributions of Table 3.2.19 are subexponential.

(12) Let (X_i) be an iid sequence of positive random variables with common distribution function F. Write $S_n = X_1 + \cdots + X_n$, $n \geq 1$.
(a) Show that for every $n \geq 1$ the following relation holds:

$$\liminf_{x \to \infty} \frac{P(S_n > x)}{n\,\overline{F}(x)} = 1.$$

(b) Show that the definition of a subexponential distribution function F is equivalent to the following relation

$$\limsup_{x \to \infty} \frac{P(S_n > x)}{P(X_i > x \text{ for some } i \leq n \text{ and } X_j \leq x \text{ for } 1 \leq j \neq i \leq n)} = 1,$$

for all $n \geq 2$.
(c) Show that for a subexponential distribution function F and $1 \leq k \leq n$,

$$\lim_{x \to \infty} P(X_1 + \cdots + X_k > x \mid X_1 + \cdots + X_n > x) = \frac{k}{n}.$$

(d) The relation (3.2.19) can be shown to hold uniformly on bounded y-intervals for subexponential F. Use this information to show that

$$\lim_{x \to \infty} P(X_1 \leq z \mid X_1 + X_2 > x) = 0.5\, F(z), \quad z > 0.$$

3.3 The Distribution of the Total Claim Amount

In this section we study the distribution of the total claim amount

$$S(t) = \sum_{i=1}^{N(t)} X_i$$

under the standard assumption that the claim number process N and the iid sequence (X_i) of positive claims are independent. We often consider the case of fixed t, i.e., we study the *random variable $S(t)$, not the stochastic process* $(S(t))_{t \geq 0}$. When t is fixed, we will often suppress the dependence of $N(t)$ and $S(t)$ on t and write $N = N(t)$, $S = S(t)$ and

$$S = \sum_{i=1}^{N} X_i \,,$$

thereby abusing our previous notation since we have used the symbols N for the claim number process and S for the total claim amount process before. It will, however, be clear from the context what S and N denote in the different sections.

In Section 3.3.1 we investigate the distribution of the total claim amount in terms of its characteristic function. We introduce the class of *mixture distributions* which turn out to be useful for characterizing the distribution of the total claim amount, in particular for compound Poisson processes. The most important results of this section say that sums of independent compound Poisson variables are again compound Poisson. Moreover, given a compound Poisson process (such as the total claim amount process in the Cramér-Lundberg model), it can be decomposed into independent compound Poisson processes by introducing a disjoint partition of time and claim size space. These results are presented in Section 3.3.2. They are extremely useful, for example, if one is interested in the total claim amount over smaller periods of time or in the total claim amount of claim sizes assuming values in certain layers. We continue in Section 3.3.3 with a numerical procedure, the *Panjer recursion*, for determining the *exact distribution* of the total claim amount. This procedure works for integer-valued claim sizes and for a limited number of claim number distributions. In Sections 3.3.4 and 3.3.5 we consider alternative methods for determining *approximations* to the distribution of the total claim amount. They are based on the central limit theorem or Monte Carlo techniques.

3.3.1 Mixture Distributions

In this section we are interested in some theoretical properties of the distribution of $S = S(t)$ for fixed t. The distribution of S is determined by its *characteristic function*

$$\phi_S(s) = E e^{isS}, \quad s \in \mathbb{R},$$

and we focus here on techniques based on characteristic functions. Alternatively, we could use the moment generating function

$$m_S(h) = E e^{hS}, \quad h \in (-h_0, h_0),$$

provided the latter is finite for some positive $h_0 > 0$. Indeed, m_S also determines the distribution of S. However, $m_S(h)$ is finite in some neighborhood of the origin if and only if the tail $P(S > x)$ decays exponentially fast, i.e.,

$$P(S > x) \le c e^{-\gamma x}, \quad x > 0,$$

for some positive c, γ. This assumption is not satisfied for S with the heavy-tailed claim size distributions introduced in Table 3.2.19, and therefore we prefer using characteristic functions,[4] which are defined for any random variable S.

Exploiting the independence of N and (X_i), a conditioning argument yields the following useful formula:

$$\phi_S(s) = E \left(E \left[e^{is(X_1 + \cdots + X_N)} \,\middle|\, N \right] \right)$$

$$= E \left(\left[E e^{isX_1} \right]^N \right) = E([\phi_{X_1}(s)]^N)$$

$$= E e^{N \log \phi_{X_1}(s)} = m_N(\log \phi_{X_1}(s)). \qquad (3.3.24)$$

(The problems we have mentioned with the moment generating function do not apply in this situation, since we consider m_N at the complex argument $\log \phi_{X_1}(s)$. The quantities in (3.3.24) are all bounded in absolute value by 1, since we deal with a characteristic function.) We apply this formula to two important examples: the compound Poisson case, i.e., when N has a Poisson distribution, and the compound geometric case, i.e., when N has a geometric distribution.

Example 3.3.1 (Compound Poisson sum)
Assume that N is Pois(λ) distributed for some $\lambda > 0$. Straightforward calculation yields

$$m_N(h) = e^{-\lambda(1 - e^h)}, \quad h \in \mathbb{C}.$$

---

[4] As a second alternative to characteristic functions we could use the Laplace-Stieltjes transform $\widehat{f}_S(s) = m_S(-s)$ for $s > 0$ which is well-defined for non-negative random variables S and determines the distribution of S. The reader who feels uncomfortable with the notion of characteristic functions could switch to moment generating functions or Laplace-Stieltjes transforms; most of the calculations can easily be adapted to either of the two transforms. We refer to p. 182 for a brief introduction to Laplace-Stieltjes transforms.

Then we conclude from (3.3.24) that

$$\phi_S(s) = e^{-\lambda(1-\phi_{X_1}(s))}, \quad s \in \mathbb{R}.$$

$\square$

Example 3.3.2 (Compound geometric sum)
We assume that N has a geometric distribution with parameter $p \in (0,1)$, i.e.,

$$P(N = n) = p\,q^n, \quad n = 0,1,2,\dots, \quad \text{where } q = 1 - p.$$

Moreover, let X_1 be exponentially $\mathrm{Exp}(\lambda)$ distributed. It is not difficult to verify that

$$\phi_{X_1}(s) = \frac{\lambda}{\lambda - is}, \quad s \in \mathbb{R}.$$

We also have

$$m_N(h) = \sum_{n=0}^{\infty} e^{nh} P(N = n) = \sum_{n=0}^{\infty} e^{nh} p\,q^n = \frac{p}{1 - e^h q}$$

provided $|h| < -\log q$. Plugging ϕ_{X_1} and m_N in formula (3.3.24), we obtain

$$\phi_S(t) = \frac{p}{1 - \lambda(\lambda - is)^{-1} q} = p + q\,\frac{\lambda p}{\lambda p - is}, \quad s \in \mathbb{R}.$$

We want to interpret the right-hand side in a particular way. Let J be a random variable assuming two values with probabilities p and q, respectively. For example, choose $P(J = 1) = p$ and $P(J = 2) = q$. Consider the random variable

$$S' = I_{\{J=1\}}\,0 + I_{\{J=2\}}\,Y,$$

where Y is $\mathrm{Exp}(\lambda p)$ distributed and independent of J. This means that we choose either the random variable 0 or the random variable Y according as $J = 1$ or $J = 2$. Writing F_A for the distribution function of any random variable A, we see that S' has distribution function

$$F_{S'}(x) = p\,F_0(x) + q\,F_Y(x) = p\,I_{[0,\infty)}(x) + q\,F_Y(x), \quad x \in \mathbb{R},$$

$$(3.3.25)$$

and characteristic function

$$Ee^{is\,S'} = P(J = 1)\,Ee^{is\,0} + P(J = 2)\,Ee^{is\,Y} = p + q\,\frac{\lambda p}{\lambda p - is}, \quad s \in \mathbb{R}.$$

In words, this is the characteristic function of S, and therefore $S \overset{d}{=} S'$:

$$S \overset{d}{=} I_{\{J=1\}}\,0 + I_{\{J=2\}}\,Y.$$

A distribution function of the type (3.3.25) determines a *mixture distribution*.

$\square$

We fix this notion in the following definition.

Definition 3.3.3 (Mixture distribution)
Let $(p_i)_{i=1,\ldots,n}$ be a distribution on the integers $\{1,\ldots,n\}$ and F_i, $i = 1,\ldots,n$, be distribution functions of real-valued random variables. Then the distribution function

$$G(x) = p_1 F_1(x) + \cdots + p_n F_n(x), \quad x \in \mathbb{R}, \tag{3.3.26}$$

defines a mixture distribution of $F_1,\ldots,F_n$.

The above definition of mixture distribution can immediately be extended to distributions (p_i) on $\{1,2,\ldots\}$ and a sequence (F_i) of distribution functions by defining

$$G(x) = \sum_{i=1}^{\infty} p_i F_i(x), \quad x \in \mathbb{R}.$$

For our purposes, finite mixtures are sufficient.

As in Example 3.3.2 of a compound geometric sum, we can interpret the probabilities p_i as the distribution of a discrete random variable J assuming the values i: $P(J = i) = p_i$. Moreover, assume J is independent of the random variables $Y_1,\ldots,Y_n$ with distribution functions $F_{Y_i} = F_i$. Then a conditioning argument shows that the random variable

$$Z = I_{\{J=1\}} Y_1 + \cdots + I_{\{J=n\}} Y_n$$

has the mixture distribution function

$$F_Z(x) = p_1 F_{Y_1}(x) + \cdots + p_n F_{Y_n}(x), \quad x \in \mathbb{R},$$

with the corresponding characteristic function

$$\phi_Z(s) = p_1 \phi_{Y_1}(s) + \cdots + p_n \phi_{Y_n}(s), \quad s \in \mathbb{R}. \tag{3.3.27}$$

It is interesting to observe that the dependence structure of the Y_i's does not matter here.

An interesting result in the context of mixture distributions is the following.

Proposition 3.3.4 (Sums of independent compound Poisson variables are compound Poisson)
Consider the independent compound Poisson sums

$$S_i = \sum_{j=1}^{N_i} X_j^{(i)}, \quad i = 1,\ldots,n,$$

where N_i is $\mathrm{Pois}(\lambda_i)$ distributed for some $\lambda_i > 0$ and $(X_j^{(i)})$ is an iid sequence of claim sizes. Then the sum

$$\widetilde{S} = S_1 + \cdots + S_n$$

is again compound Poisson with representation

$$\widetilde{S} \stackrel{d}{=} \sum_{i=1}^{N_\lambda} Y_i , \quad N_\lambda \sim \text{Pois}(\lambda) , \quad \lambda = \lambda_1 + \cdots + \lambda_n ,$$

and (Y_i) is an iid sequence, independent of N_λ, with mixture distribution (3.3.26) given by

$$p_i = \lambda_i/\lambda \quad and \quad F_i = F_{X_1^{(i)}} . \tag{3.3.28}$$

Proof. Recall the characteristic function of a compound Poisson variable from Example 3.3.1:

$$\phi_{S_j}(s) = \exp\left\{ -\lambda_j \left(1 - \phi_{X_1^{(j)}}(s) \right) \right\} , \quad s \in \mathbb{R} .$$

By independence of the S_j's and the definition (3.3.28) of the p_j's,

$$\phi_{\widetilde{S}}(s) = \phi_{S_1}(s) \cdots \phi_{S_n}(s)$$

$$= \exp\left\{ -\lambda \sum_{j=1}^{n} p_j \left(1 - \phi_{X_1^{(j)}}(s) \right) \right\}$$

$$= \exp\left\{ -\lambda \left(1 - E \exp\left\{ is \sum_{j=1}^{n} I_{\{J=j\}} X_1^{(j)} \right\} \right) \right\} , \quad s \in \mathbb{R} ,$$

where J is independent of the $X_1^{(j)}$'s and has distribution $(P(J = i))_{i=1,\ldots,n} = (p_i)_{i=1,\ldots,n}$. This is the characteristic function of a compound Poisson sum with summands whose distribution is described in (3.3.27), where (p_i) and (F_i) are specified in (3.3.28).   □

The fact that sums of independent compound Poisson random variables are again compound Poisson is a nice closure property which has interesting applications in insurance. We illustrate this in the following example.

Example 3.3.5 (Applications of the compound Poisson property)
(1) Consider a Poisson process $N = (N(t))_{t \geq 0}$ with mean value function μ and assume that the claim sizes in the portfolio in year i constitute an iid sequence $(X_j^{(i)})$ and that all sequences $(X_j^{(i)})$ are mutually independent and independent of the claim number process N. The total claim amount in year i is given by

$$S_i = \sum_{j=N(i-1)+1}^{N(i)} X_j^{(i)} .$$

Since N has independent increments and the iid sequences $(X_j^{(i)})$ are mutually independent, we observe that

$$\left(\sum_{j=N(i-1)+1}^{N(i)} X_j^{(i)}\right)_{i=1,\dots,n} \overset{d}{=} \left(\sum_{j=1}^{N(i-1,i]} X_j^{(i)}\right)_{i=1,\dots,n}. \qquad (3.3.29)$$

A formal proof of this identity is easily provided by identifying the joint characteristic functions of the vectors on both sides. This verification is left as an exercise. Since $(N(i-1,i])$ is a sequence of independent random variables, independent of the independent sequences $(X_j^{(i)})$, the annual total claim amounts S_i are mutually independent. Moreover, each of them is compound Poisson: let N_i be $\text{Pois}(\mu(i-1,i])$ distributed, independent of $(X_j^{(i)})$, $i=1,\dots,n$. Then

$$S_i \overset{d}{=} \sum_{j=1}^{N_i} X_j^{(i)}.$$

We may conclude from Proposition 3.3.4 that the total claim amount $S(n)$ in the first n years is again compound Poisson, i.e.,

$$S(n) = S_1 + \cdots + S_n \overset{d}{=} \sum_{i=1}^{N_\lambda} Y_i,$$

where the random variable

$$N_\lambda \sim \text{Pois}(\lambda), \quad \lambda = \mu(0,1] + \cdots + \mu(n-1,n] = \mu(n),$$

is independent of the iid sequence (Y_i). Each of the Y_i's has representation

$$Y_i \overset{d}{=} I_{\{J=1\}} X_1^{(1)} + \cdots + I_{\{J=n\}} X_1^{(n)}, \qquad (3.3.30)$$

where J is independent of the $X_1^{(j)}$'s, with distribution $P(J = i) = \mu(i-1,i]/\lambda$.

In other words, the total claim amount $S(n)$ in the first n years with possibly different claim size distributions in each year has representation as a compound Poisson sum with Poisson counting variable N_λ which has the same distribution as $N(n)$ and with iid claim sizes Y_i with the mixture distribution presented in (3.3.30).

(2) Consider n independent portfolios with total claim amounts in a fixed period of time given by the compound Poisson sums

$$S_i = \sum_{j=1}^{N_i} X_j^{(i)}, \quad N_i \sim \text{Pois}(\lambda_i).$$

The claim sizes $X_j^{(i)}$ in the ith portfolio are iid, but the distributions may differ from portfolio to portfolio. For example, think of each portfolio as a collection of policies corresponding to one particular type of car insurance or, even simpler, think of each portfolio as the claim history in one particular policy. Now, Proposition 3.3.4 ensures that the aggregation of the total claim amounts from the different portfolios, i.e.,

$$\widetilde{S} = S_1 + \cdots + S_n \,,$$

is again compound Poisson with counting variable which has the same Poisson distribution as $N_1 + \cdots + N_n \sim \text{Pois}(\lambda)$, $\lambda = \lambda_1 + \cdots + \lambda_n$, with iid claim sizes Y_i. A sequence of the Y_i's can be realized by independent repetitions of the following procedure:

(a) Draw a number $i \in \{1, \ldots, n\}$ with probability $p_i = \lambda_i / \lambda$.
(b) Draw a realization from the claim size distribution of the ith portfolio.

$\square$

3.3.2 Space-Time Decomposition of a Compound Poisson Process

In this section we prove a converse result to Proposition 3.3.4: we decompose a compound Poisson process into independent compound Poisson processes by partitioning time and (claim size) space. In this context, we consider a general compound Poisson process

$$S(t) = \sum_{i=1}^{N(t)} X_i \,, \quad t \geq 0 \,,$$

where N is a Poisson process on $[0, \infty)$ with mean value function μ and arrival sequence (T_i), independent of the iid sequence (X_i) of positive claim sizes of common distribution F. The mean value function μ generates a measure on the Borel σ-field of $[0, \infty)$, the mean measure of the Poisson process N, which we also denote by μ.

The points (T_i, X_i) assume values in the state space $E = [0, \infty)^2$ equipped with the Borel σ-field $\mathcal{E}$. We have learned in Section 2.1.8 that the counting measure

$$M(A) = \{i \geq 1 : (T_i, X_i) \in A\} \,, \quad A \in \mathcal{E} \,,$$

is a Poisson random measure with mean measure $\nu = \mu \times F$. This means in particular that for any disjoint partition $A_1, \ldots, A_n$ of E, i.e.,

$$\bigcup_{i=1}^{n} A_i = E \,, \quad A_i \cap A_j = \emptyset \,, \quad 1 \leq i < j \leq n \,,$$

the random variables $M(A_1), \ldots, M(A_n)$ are independent and $M(A_i) \sim \text{Pois}(\nu(A_i))$, $i = 1, \ldots, n$, where we interpret $M(A_i) = \infty$ if $\nu(A_i) = \infty$. But even more is true, as the following theorem shows:

Theorem 3.3.6 (Space-time decomposition of a compound Poisson sum)
Assume that the mean value function μ of the Poisson process N on $[0, \infty)$ has an a.e. positive continuous intensity function λ. Let $A_1, \ldots, A_n$ be a disjoint partition of $E = [0, \infty)^2$. Then the following statements hold.

(1) *For every $t \geq 0$, the random variables*

$$S_j(t) = \sum_{i=1}^{N(t)} X_i \, I_{A_j}((T_i, X_i)), \quad j = 1, \ldots, n,$$

are mutually independent.

(2) *For every $t \geq 0$, $S_j(t)$ has representation as a compound Poisson sum*

$$S_j(t) \stackrel{d}{=} \sum_{i=1}^{N(t)} X_i \, I_{A_j}((Y_i, X_i)), \tag{3.3.31}$$

where (Y_i) is an iid sequence of random variables with density $\lambda(x)/\mu(t)$ $0 \leq x \leq t$, independent of N and (X_i).

Proof. Since μ has an a.e. positive continuous intensity function λ we know from the order statistics property of the one-dimensional Poisson process N (see Theorem 2.1.11) that

$$(T_1, \ldots, T_k \mid N(t) = k) \stackrel{d}{=} (Y_{(1)}, \ldots, Y_{(k)}),$$

where $Y_{(1)} \leq \cdots \leq Y_{(k)}$ are the order statistics of an iid sample $Y_1, \ldots, Y_k$ with common density $\lambda(x)/\mu(t)$, $0 \leq x \leq t$. By a similar argument as in the proof of Proposition 2.1.16 we may conclude that

$$((S_j(t))_{j=1,\ldots,n} \mid N(t) = k) \tag{3.3.32}$$

$$\stackrel{d}{=} \left(\sum_{i=1}^{k} X_i \, I_{A_j}((Y_{(i)}, X_i)) \right)_{j=1,\ldots,n} \stackrel{d}{=} \left(\sum_{i=1}^{k} X_i \, I_{A_j}((Y_i, X_i)) \right)_{j=1,\ldots,n},$$

where N, (Y_i) and (X_i) are independent. Observe that each of the sums on the right-hand side has iid summands. We consider the joint characteristic function of the $S_j(t)$'s. Exploiting relation (3.3.32), we obtain for any $s_i \in \mathbb{R}$, $i = 1, \ldots, n$,

$$\phi_{S_1(t),\ldots,S_n(t)}(s_1, \ldots, s_n)$$

$$= E e^{i s_1 S_1(t) + \cdots + i s_n S_n(t)}$$

$$= \sum_{k=0}^{\infty} P(N(t) = k) \, E \left(e^{i s_1 S_1(t) + \cdots + i s_n S_n(t)} \,\Big|\, N(t) = k \right)$$

$$= \sum_{k=0}^{\infty} P(N(t) = k) \, E \exp \left\{ i \sum_{l=1}^{k} \sum_{j=1}^{n} s_j \, X_l \, I_{A_j}((Y_l, X_l)) \right\}$$

$$= E \exp \left\{ i \sum_{l=1}^{N(t)} \sum_{j=1}^{n} s_j \, X_l \, I_{A_j}((Y_l, X_l)) \right\} .$$

Notice that the exponent in the last line is a compound Poisson sum. From the familiar form of its characteristic function and the disjointness of the A_j's we may conclude that

$$\log \phi_{S_1(t), \ldots, S_n(t)}(s_1, \ldots, s_n)$$

$$= -\mu(t) \left(1 - E \exp \left\{ i \sum_{j=1}^{n} s_j \, X_1 \, I_{A_j}((Y_1, X_1)) \right\} \right)$$

$$= -\mu(t) \left(1 - \left[\sum_{j=1}^{n} \left(E e^{\, i \, s_j \, X_1 \, I_{A_j}((Y_1, X_1))} - \left(1 - P((Y_1, X_1) \in A_j) \right) \right) \right] \right)$$

$$= -\mu(t) \sum_{j=1}^{n} \left(1 - E e^{\, i \, s_j \, X_1 \, I_{A_j}((Y_1, X_1))} \right) . \tag{3.3.33}$$

The right-hand side in (3.3.33) is nothing but the sum of the logarithms of the characteristic functions $\phi_{S_j(t)}(s_j)$. Equivalently, the joint characteristic function of the $S_j(t)$'s factorizes into the individual characteristic functions $\phi_{S_j(t)}(s_j)$. This means that the random variables $S_j(t)$ are mutually independent and each of them has compound Poisson structure as described in (3.3.31), where we again used the identity in law (3.3.32). This proves the theorem. □

Theorem 3.3.6 has a number of interesting consequences.

Example 3.3.7 (Decomposition of time and claim size space in the Cramér-Lundberg model)
Consider the total claim amount process S in the Cramér-Lundberg model with Poisson intensity $\lambda > 0$ and claim size distribution function F.

(1) *Partitioning time.* Choose $0 = t_0 < t_1 < \ldots < t_n = t$ and write

$$\Delta_1 = [0, t_1], \quad \Delta_i = (t_{i-1}, t_i], \quad i = 2, \ldots, n, \quad \Delta_{n+1} = (t_n, \infty).$$
$$\tag{3.3.34}$$

Then

$$A_i = \Delta_i \times [0, \infty), \quad i = 1, \ldots, n+1,$$

is a disjoint decomposition of the state space $E = [0, \infty)^2$. An application of Theorem 3.3.6 yields that the random variables

$$\sum_{i=1}^{N(t)} X_i\, I_{A_j}((T_i, X_i)) = \sum_{i=N(t_{j-1})+1}^{N(t_j)} X_i\,, \quad j = 1, \ldots, n\,,$$

are independent. This is the well-known independent increment property of the compound Poisson process. It is also not difficult to see that the increments are stationary, i.e., $S(t) - S(s) \stackrel{d}{=} S(t - s)$ for $s < t$. Hence they are again compound Poisson sums.

(2) *Partitioning claim size space.* For fixed t, we partition the claim size space $[0, \infty)$ into the disjoint sets $B_0, \ldots, B_n$. For example, one can think of disjoint *layers*

$$B_1 = [0, d_1]\,, B_2 = (d_1, d_2]\,, \ldots, B_n = (d_{n-1}, d_n]\,, B_{n+1} = (d_n, \infty)\,,$$

where $0 < d_1 < \cdots < d_n < \infty$ are finitely many limits which classify the order of magnitude of the claim sizes. Such layers are considered in a reinsurance context, where different insurance companies share the risk (and the premium) of a portfolio in its distinct layers. Then the sets

$$A_i = [0, t] \times B_i\,, \quad A_i' = (t, \infty) \times B_i\,, \quad i = 1, \ldots, n+1\,,$$

constitute a disjoint partition of the state space E. An application of Theorem 3.3.6 yields that the total claim amounts in the different parts of the partition

$$S_j(t) = \sum_{i=1}^{N(t)} X_i\, I_{A_j}((T_i, X_i)) = \sum_{i=1}^{N(t)} X_i I_{B_j}(X_i)\,, \quad j = 1, \ldots, n+1\,,$$

are mutually independent. Whereas the independent increment property of S is perhaps not totally unexpected because of the corresponding property of the Poisson process N, the independence of the quantities $S_j(t)$ is not obvious from their construction. Their compound Poisson structure is, however, immediate since the summands $X_i I_{B_j}(X_i)$ are iid and independent of $N(t)$.

(3) *General partitions.* So far we partitioned either time or the claim size space. But Theorem 3.3.6 allows one to consider *any* disjoint partition of the state space E. The message is always the same: the total claim amounts on the distinct parts of the partition are independent and have compound Poisson structure. This is an amazing and very useful result.   □

Example 3.3.8 (Partitioning claim size space and time in an IBNR portfolio)
Let (T_i) be the claim arrival sequence of a Poisson process N on $[0, \infty)$ with mean value function μ, independent of the sequence (V_i) of iid positive delay random variables with distribution function F. Consider a sequence (X_i)

of iid positive claim sizes, independent of (T_i) and (V_i). We have learned in Example 2.1.29 that the points $(T_i + V_i)$ of the reporting times of the claims constitute a Poisson process (PRM) N_{IBNR} with mean value function $\nu(t) = \int_0^t F(t-s)\,d\mu(s)$. The total claim amount $S(t)$ in such an IBNR portfolio, i.e., the total claim amount of the claim sizes which are reported by time t, is described by

$$S(t) = \sum_{i=1}^{N_{\mathrm{IBNR}}(t)} X_i, \quad t \geq 0.$$

Theorem 3.3.6 now ensures that we can split time and/or claim size space in the same way as for the total claim amount in the Cramér-Lundberg model, i.e., the total claim amounts in the different parts of the partition constitute independent compound Poisson sums. The calculations are similar to Example 3.3.7; we omit further details. $\qquad\square$

Theorem 3.3.6 has immediate consequences for the dependence structure of the compound Poisson processes of the decomposition of the total claim amount.

Corollary 3.3.9 *Under the conditions of Theorem 3.3.6, the processes $S_j = (S_j(t))_{t \geq 0}$, $j = 1, \ldots, n$ are mutually independent and have independent increments.*

Proof. We start by showing the independent increment property for one process S_j. For $0 = t_0 < \cdots < t_n$ and $n \geq 1$, define the Δ_i's as in (3.3.34). The sets

$$A_i' = A_i \cap (\Delta_i \times [0, \infty)), \quad i = 1, \ldots, n,$$

are disjoint. An application of Theorem 3.3.6 yields that the random variables

$$\sum_{i=1}^{N(t_n)} X_i\, I_{A_j'}((T_i, X_i)) = \sum_{i=N(t_{j-1})+1}^{N(t_j)} X_i\, I_{A_j}((T_i, X_i)) = S_j(t_{i-1}, t_i]$$

are mutually independent. This means that the process S_j has independent increments.

In order to show the independence of the processes S_j, $j = 1, \ldots, n$, one has to show that the families of the random variables $(S_j(t_i^{(j)}))_{i=1,\ldots,k_j}$, $j = 1, \ldots, n$ for any choices of increasing $t_i^{(j)} \geq 0$ and integers $k_j \geq 1$ are mutually independent. Define the quantities $\Delta_i^{(j)}$ for $0 = t_0^{(j)} < \cdots < t_{k_j}^{(j)} < \infty$, $j = 1, \ldots, n$, in analogy to (3.3.34). Then

$$A_i^{(j)} = A_j \cap \left(\Delta_i^{(j)} \times [0, \infty)\right), \quad i = 1, \ldots, k_j, \quad j = 1, \ldots, n,$$

are disjoint subsets of E. By the same argument as above, the increments

$$S_j(t_{i-1}^{(j)}, t_i^{(j)}], \quad i = 1, \ldots, k_j, \quad j = 1, \ldots, n,$$

are independent. We conclude that the families of the random variables

$$\left(S_j(t_i^{(j)})\right)_{i=1,\ldots,k_j} = \left(\sum_{k=1}^{i} S_j(t_{k-1}^{(j)}, t_k^{(j)}]\right)_{i=1,\ldots,k_j}, \quad j = 1, \ldots, n,$$

are mutually independent: for each j, the $S_j(t_i^{(j)})$'s are constructed from increments which are mutually independent of the increments of S_k, $k \neq j$. $\square$

3.3.3 An Exact Numerical Procedure for Calculating the Total Claim Amount Distribution

In this section we consider one particular exact numerical technique which has become popular in insurance practice. As in Section 3.3.1, we consider $S(t)$ for fixed t, and therefore we suppress the dependence of $S(t)$ and $N(t)$ on t, i.e., we write

$$S = \sum_{i=1}^{N} X_i$$

for an integer-valued random variable N, independent of the iid claim size sequence (X_i). We also write

$$S_0 = 0, \quad S_n = X_1 + \cdots + X_n, \quad n \geq 1,$$

for the partial sum process (random walk) generated by the claim sizes X_i.

The distribution function of S is given by

$$P(S \leq x) = E[P(S \leq x \mid N)] = \sum_{n=0}^{\infty} P(S_n \leq x) P(N = n).$$

From this formula we see that the total claim amount S has quite a complicated structure: even if we knew the probabilities $P(N = n)$ and the distribution of X_i, we would have to calculate the distribution functions of all partial sums S_n. This mission is impossible, in general. In general, we can say little about the exact distribution of S, and so one is forced to use Monte Carlo or numerical techniques for approximating the total claim amount distribution.

The numerical method we focus on yields the *exact* distribution of the total claim amount S. This procedure is often referred to as *Panjer recursion*, since its basic idea goes back to Harry Panjer [60]. The method is restricted to claim size distributions with support on a lattice (such as the integers) and to a limited class of claim number distributions. By now, high speed computers with a huge memory allow for efficient alternative Monte Carlo and numerical procedures in more general situations.

We start by giving the basic assumptions under which the method works.

(1) The claim sizes X_i assume values in $\mathbb{N}_0 = \{0, 1, 2, \ldots\}$.
(2) The claim number N has distribution of type

$$q_n = P(N = n) = \left(a + \frac{b}{n}\right) q_{n-1}, \quad n = 1, 2, \ldots,$$

for some $a, b \in \mathbb{R}$.

Condition (1) is slightly more general than it seems. Alternatively, one could assume that X_i assumes values in the lattice $d\mathbb{N}_0$ for some $d > 0$. Indeed, we then have $S = d \sum_{i=1}^{N}(X_i/d)$, and the random variables X_i/d assume values in $\mathbb{N}_0$.

Condition (1) rules out all continuous claim size distributions, in particular, those with a density. One might argue that this is not really a restriction since

(a) every continuous claim size distribution on $[0, \infty)$ can be approximated by a lattice distribution arbitrarily closely (for example, in the sense of uniform or total variation distance) if one chooses the span of the lattice sufficiently small,
(b) all real-life claim sizes are expressed in terms of prices which, necessarily, take values on a lattice.

Note, however, that fact (a) does not give any information about the goodness of the approximation to the distribution of S, if the continuous claim size distribution is approximated by a distribution on a lattice. As regards (b), observe that all claim size distributions which have been relevant in the history of insurance mathematics (see Tables 3.2.17 and 3.2.19) have a density and would therefore fall outside the considerations of the present section.

Condition (2) is often referred to as (a, b)-condition. It is not difficult to verify that three standard claim number distributions satisfy this condition:

(a) The Poisson $\mathrm{Pois}(\lambda)$ distribution[5] with $a = 0$, $b = \lambda \geq 0$. In this case one obtains the (a, b)-region $R_{\mathrm{Pois}} = \{(a, b) : a = 0, b \geq 0\}$.
(b) The binomial $\mathrm{Bin}(n, p)$ distribution[6] with $a = -p/(1-p) < 0$, $b = -a(n+1)$, $n \geq 0$. In this case one obtains the (a, b)-region $R_{\mathrm{Bin}} = \{(a, b) : a < 0, b = -a(n+1) \text{ for some integer } n \geq 0\}$.
(c) The negative binomial distribution with parameters (p, v), see Example 2.3.3, with $0 < a = 1 - p < 1$, $b = (1 - p)(v - 1)$ and $a + b > 0$. In this case one obtains the (a, b)-region $R_{\mathrm{Negbin}} = \{(a, b) : 0 < a < 1, a+b > 0\}$.

These three distributions are the only distributions on $\mathbb{N}_0$ satisfying the (a, b)-condition. In particular, only for the (a, b)-parameter regions indicated above the (a, b)-condition yields genuine distributions (q_n) on $\mathbb{N}_0$. The verification of these statements is left as an exercise; see Exercise 7 on p. 145.

Now we formulate the *Panjer recursion scheme*.

---

[5] The case $\lambda = 0$ corresponds to the distribution of $N = 0$.
[6] The case $n = 0$ corresponds to the distribution of $N = 0$.

Theorem 3.3.10 (Panjer recursion scheme)
Assume conditions (1) *and* (2) *on the distributions of* X_i *and* N. *Then the probabilities* $p_n = P(S = n)$ *can be calculated recursively as follows:*

$$p_0 = \begin{cases} q_0 & \text{if } P(X_1 = 0) = 0, \\ E([P(X_1 = 0)]^N) & \text{otherwise.} \end{cases}$$

$$p_n = \frac{1}{1 - a\,P(X_1 = 0)} \sum_{i=1}^{n} \left(a + \frac{b\,i}{n} \right) P(X_1 = i)\, p_{n-i}, \quad n \geq 1.$$

Since the parameter a is necessarily less than 1, all formulae for p_n are well-defined.

Proof. We start with

$$p_0 = P(N = 0) + P(S = 0, N > 0).$$

The latter relation equals q_0 if $P(X_1 = 0) = 0$. Otherwise,

$$p_0 = q_0 + \sum_{i=1}^{\infty} P(X_1 = 0, \dots, X_i = 0)\, P(N = i)$$

$$= q_0 + \sum_{i=1}^{\infty} [P(X_1 = 0)]^i\, P(N = i)$$

$$= E([P(X_1 = 0)]^N).$$

Now we turn to the case p_n, $n \geq 1$. A conditioning argument and the (a, b)-condition yield

$$p_n = \sum_{i=1}^{\infty} P(S_i = n)\, q_i = \sum_{i=1}^{\infty} P(S_i = n) \left(a + \frac{b}{i} \right) q_{i-1}. \qquad (3.3.35)$$

Notice that

$$E\left(a + \frac{b\,X_1}{n} \,\Big|\, S_i = n \right) = E\left(a + \frac{b\,X_1}{X_1 + \dots + X_i} \,\Big|\, S_i = n \right)$$

$$= a + \frac{b}{i}, \qquad (3.3.36)$$

since by the iid property of the X_i's

$$1 = E\left(\frac{S_i}{S_i} \,\Big|\, S_i \right) = \sum_{k=1}^{i} E\left(\frac{X_k}{S_i} \,\Big|\, S_i \right) = i\, E\left(\frac{X_1}{S_i} \,\Big|\, S_i \right).$$

We also observe that

$$E\left(a + \frac{b\,X_1}{n}\,\bigg|\,S_i = n\right)$$

$$= \sum_{k=0}^{n}\left(a + \frac{b\,k}{n}\right) P(X_1 = k \mid S_i = n)$$

$$= \sum_{k=0}^{n}\left(a + \frac{b\,k}{n}\right) \frac{P(X_1 = k\,,\,S_i - X_1 = n - k)}{P(S_i = n)}$$

$$= \sum_{k=0}^{n}\left(a + \frac{b\,k}{n}\right) \frac{P(X_1 = k)\,P(S_{i-1} = n - k)}{P(S_i = n)}\,. \qquad (3.3.37)$$

Substitute (3.3.36) and (3.3.37) into (3.3.35) and interchange the order of summation:

$$p_n = \sum_{i=1}^{\infty}\sum_{k=0}^{n}\left(a + \frac{b\,k}{n}\right) P(X_1 = k)\,P(S_{i-1} = n - k)\,q_{i-1}$$

$$= \sum_{k=0}^{n}\left(a + \frac{b\,k}{n}\right) P(X_1 = k)\left[\sum_{i=1}^{\infty} P(S_{i-1} = n - k)\,q_{i-1}\right]$$

$$= \sum_{k=0}^{n}\left(a + \frac{b\,k}{n}\right) P(X_1 = k)\,P(S = n - k)$$

$$= \sum_{k=0}^{n}\left(a + \frac{b\,k}{n}\right) P(X_1 = k)\,p_{n-k}\,.$$

Thus we finally obtain

$$p_n = a\,P(X_1 = 0)\,p_n + \sum_{k=1}^{n}\left(a + \frac{b\,k}{n}\right) P(X_1 = k)\,p_{n-k}\,,$$

which gives the final result for p_n. □

Example 3.3.11 (Stop-loss reinsurance contract)
We consider a so-called *stop-loss* reinsurance contract with retention level s; see also Section 3.4. This means that the reinsurer covers the excess $(S - s)_+$ of the total claim amount S over the threshold s. Suppose the company is interested in its net premium, i.e., the expected loss:

$$p(s) = E(S - s)_+ = \int_{s}^{\infty} P(S > x)\,dx\,.$$

Now assume that S is integer-valued and $s \in \mathbb{N}_0$. Then

$$p(s) = \sum_{k=s}^{\infty} P(S > k) = p(s-1) - P(S > s-1).$$

This yields a recursive relation for $p(s)$:

$$p(s) = p(s-1) - [1 - P(S \le s-1)].$$

The probability $P(S \le s-1) = \sum_{i=0}^{s-1} p_i$ can be calculated by Panjer recursion from $p_1, \ldots, p_{s-1}$. Now, starting with the initial value $p(0) = ES = EN\, EX_1$, we have a recursive scheme for calculating the net premium of a stop-loss contract. □

Comments

Papers on extensions of Panjer's recursion have frequently appeared in the journal *ASTIN Bulletin*. The interested reader is referred, for example, to Sundt [77] or Hess et al. [43]. The book by Kaas et al. [46] contains a variety of numerical methods for the approximation of the total claim amount distribution and examples illustrating them. See also the book by Willmot and Lin [80] on approximations to compound distributions. The monographs by Asmussen [4] and Rolski et al. [67] contain chapters about the approximation of the total claim amount distribution.

The following papers on the computation of compound sum distributions can be highly recommended: Grübel and Hermesmeier [38, 39] and Embrechts et al. [28]. These papers discuss the use of transform methods such as the Fast Fourier Transform (FFT) for computing the distribution of compound sums as well as the discretization error one encounters when a claim size distribution is replaced by a distribution on a lattice. Embrechts et al. [28] give some basic theoretical results. Grübel and Hermesmeier [38] discuss the so-called *aliasing error* which occurs in transform methods. In recursion and transform methods one has to truncate the calculation at a level n, say. This means that one calculates a finite number of probabilities $p_0, p_1, \ldots, p_n$, where $p_k = P(S = k)$. With recursion methods one can calculate these probabilities *in principle without error*.[7] In transform methods an additional aliasing error is introduced which is essentially a wraparound effect due to the replacement of the usual summation of the integers by summation modulo the truncation point n. However, it is shown in [38] that the complexity of the FFT method is of the order $n \log n$, i.e., one needs an operation count (number of multiplications) of this order. Recursion methods require an operation count of the order n^2. With respect to this criterion, transform methods clearly outperform recursion methods. Grübel and Hermesmeier [39] also suggest an extrapolation method in order to reduce the discretization error when continuous distributions are replaced by distributions on a lattice, and they also give bounds for the discretization error.

---

[7] There is, of course, an error one encounters from floating point representations of the numbers by the computer.

3.3.4 Approximation to the Distribution of the Total Claim Amount Using the Central Limit Theorem

In this section we consider some approximation techniques for the total claim amount based on the central limit theorem. This is in contrast to Section 3.3.3, where one could determine the exact probabilities $P(S(t) = n)$ for integer-valued $S(t)$ and distributions of $N(t)$ which are in the (a, b)-class. The latter two restrictions are not needed in this section.

In our notation we switch back to the time dependent total claim amount process $S = (S(t))_{t \geq 0}$. Throughout we assume the renewal model

$$S(t) = \sum_{i=1}^{N(t)} X_i, \quad t \geq 0,$$

where the iid sequence (X_i) of positive claim sizes is independent of the renewal process $N = (N(t))_{t \geq 0}$ with arrival times $0 < T_1 < T_2 < \cdots$; see Section 2.2. Denoting the iid positive inter-arrival times as usual by $W_n = T_n - T_{n-1}$ and $T_0 = 0$, we learned in Theorem 3.1.5 about the central limit theorem for S: if $\text{var}(W_1) < \infty$ and $\text{var}(X_1) < \infty$, then

$$\sup_{x \in \mathbb{R}} \left| P\left(\frac{S(t) - ES(t)}{\sqrt{\text{var}(S(t))}} \leq x \right) - \Phi(x) \right| \tag{3.3.38}$$

$$= \sup_{y \in \mathbb{R}} \left| P(S(t) \leq y) - \Phi((y - ES(t))/\sqrt{\text{var}(S(t))}) \right| \to 0, \tag{3.3.39}$$

where Φ is the distribution function of the standard normal $N(0, 1)$ distribution. As in classical statistics, where one is interested in the construction of asymptotic confidence bands for estimators and in hypothesis testing, one could take this central limit theorem as justification for replacing the distribution of $S(t)$ by the normal distribution with mean $ES(t)$ and variance $\text{var}(S(t))$: for large t,

$$P(S(t) \leq y) \approx \Phi((y - ES(t))/\sqrt{\text{var}(S(t))}). \tag{3.3.40}$$

Then, for example,

$$P\left(S(t) \in [ES(t) - 1.96\sqrt{\text{var}(S(t))}, ES(t) + 1.96\sqrt{\text{var}(S(t))}] \right) \approx 0.95.$$

Relation (3.3.39) is a uniform convergence result, but it does not tell us anything about the error we encounter in (3.3.40). Moreover, when we deal with heavy-tailed claim size distributions the probability $P(S(t) > y)$ can be non-negligible even for large values of y and fixed t; see Example 3.3.13 below. The normal approximation to the tail probabilities $P(S(t) > y)$ and $P(S(t) \leq -y)$ for large y is not satisfactory (also not in the light-tailed case).

Improvements on the central limit theorem (3.3.39) have been considered starting in the 1950s. We refer to Petrov's classical monograph [61] which gives a very good overview for these kinds of results. It covers, among other things, *rates of convergence in the central limit theorem* for the partial sums

$$S_0 = 0, \quad S_n = X_1 + \cdots + X_n, \quad n \geq 1,$$

and *asymptotic expansions* for the distribution function of S_n. In the latter case, one adds more terms to $\Phi(x)$ which depend on certain moments of X_i. This construction can be shown to improve upon the normal approximation (3.3.38) substantially. The monograph by Hall [41] deals with asymptotic expansions with applications to statistics. Jensen's [45] book gives very precise approximations to probabilities of rare events (such as $P(S(t) > y)$ for values y larger than $ES(t)$), extending asymptotic expansions to *saddlepoint approximations*. Asymptotic expansions have also been derived for the distribution of the random sums $S(t)$; Chossy and Rappl [22] consider them with applications to insurance.

A rather precise tool for measuring the distance between Φ and the distribution of S_n is the so-called *Berry-Esséen inequality*. It says that

$$\sup_x (1 + |x|^3) \left| P\left(\frac{S_n - n\, EX_1}{\sqrt{n\, \mathrm{var}(X_1)}} \leq x \right) - \Phi(x) \right| \leq \frac{c}{\sqrt{n}} \frac{E|X_1 - EX_1|^3}{(\sqrt{\mathrm{var}(X_1)})^3},$$

$$(3.3.41)$$

where $c = 0.7655 + 8\,(1 + \mathrm{e}) = 30.51\ldots$ is a universal constant. Here we assumed that $E|X_1|^3 < \infty$; see Petrov [61]. The constant c can be replaced by 0.7655 if one cancels $1 + |x|^3$ on the left-hand side of (3.3.41).

Relation (3.3.41) is rather precise for various discrete distributions. For example, one can show[8] that one can derive a lower bound in (3.3.41) of the order $1/\sqrt{n}$ for iid Bernoulli random variables X_i with $P(X_i = \pm 1) = 0.5$. For distributions with a smooth density the estimate (3.3.41) is quite pessimistic, i.e., the right-hand side can often be replaced by better bounds. However, inequality (3.3.41) should be a warning to anyone who uses the central limit theorem without thinking about the error he/she encounters when the distribution of S_n is replaced by a normal distribution. It tells us that we need a sufficiently high sample size n to enable us to work with the normal distribution. But we also have to take into account the ratio $E|X_1 - EX_1|^3/(\sqrt{\mathrm{var}(X_1)})^3$, which depends on the individual distribution of X_1.

It is not possible to replace S_n by the total claim amount $S(t)$ without further work. However, we obtain a bound in the central limit theorem for $S(t)$, conditionally on $N(t) = n(t)$. Indeed, for a realization $n(t) = N(t, \omega)$ of the claim number process N we immediately have from (3.3.41) that for every $x \in \mathbb{R}$,

---

[8] Calculate the asymptotic order of the probability $P(S_{2n} = 0)$.

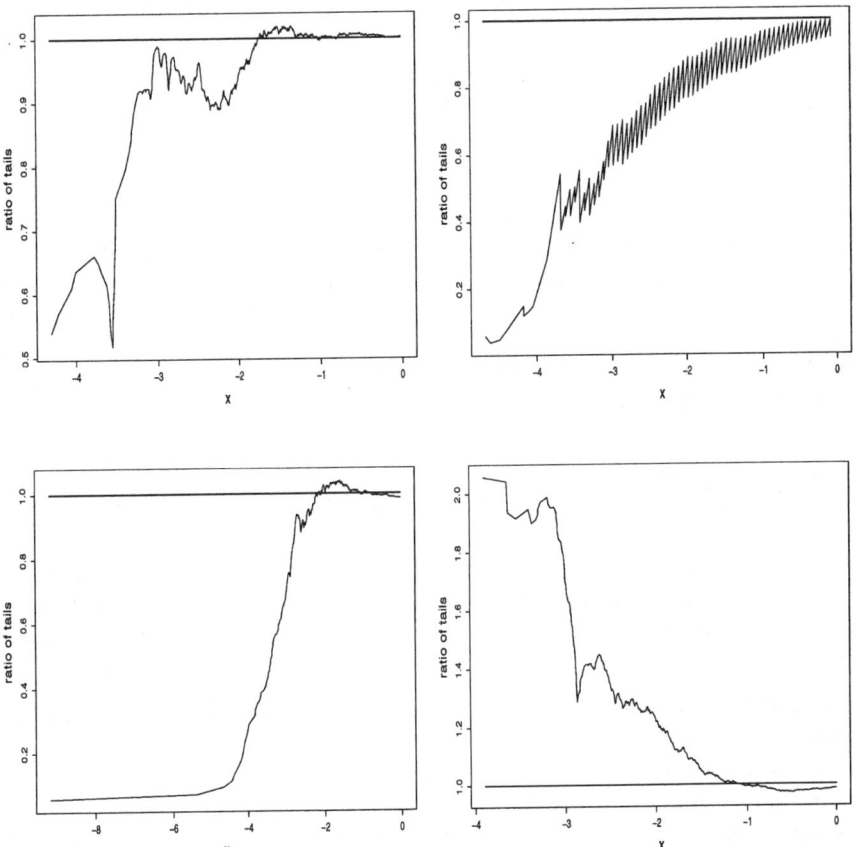

Figure 3.3.12 *A plot of the tail ratio* $r_n(x) = P((S_n - ES_n)/\sqrt{\text{var}(S_n)} \leq -x)/\Phi(-x)$, $x \geq 0$, *for the partial sums* $S_n = X_1 + \cdots + X_n$ *of iid random variables* X_i. *Here* Φ *stands for the standard normal distribution function. The order of magnitude of the deviation* $r_n(x)$ *from the constant 1 (indicated by the straight line) is a measure of the quality of the validity of the central limit theorem in the left tail of the distribution function of* S_n. *Top left:* $X_1 \sim \text{U}(0, 1)$, $n = 100$. *The central limit theorem gives a good approximation for* $x \in [-2, 0]$, *but is rather poor outside this area. Top right:* $X_1 \sim \text{Bin}(5, 0.5)$, $n = 200$. *The approximation by the central limit theorem is poor everywhere. Bottom left:* X_1 *has a* t_3-*distribution,* $n = 2\,000$. *This distribution has infinite 3rd moment and it is subexponential; cf. also Example 3.3.13. The approximation outside the area* $x \in [-3, 0]$ *is very poor due to very heavy tails of the* t_3-*distribution. Bottom right:* $X_1 \sim \text{Exp}(1)$, $n = 200$. *Although the tail of this distribution is much lighter than for the* t_3-*distribution the approximation below* $x = -1$ *is not satisfactory.*

$$\left| P\left(\frac{S(t) - n(t)EX_1}{\sqrt{n(t)\,\mathrm{var}(X_1)}} \le x \,\middle|\, N(t) = n(t) \right) - \Phi(x) \right|$$

$$\le \frac{c}{\sqrt{n(t)}} \frac{1}{1 + |x|^3} \frac{E|X_1 - EX_1|^3}{(\sqrt{\mathrm{var}(X_1)})^3}. \qquad (3.3.42)$$

Since $n(t) = N(t, \omega) \overset{\text{a.s.}}{\to} \infty$ in the renewal model, this error bound can give some justification for applying the central limit theorem to the distribution of $S(t)$, conditionally on $N(t)$, although it does not solve the original problem for the unconditional distribution of $S(t)$. In a portfolio with a large number $n(t)$ of claims, relation (3.3.42) tells us that the central limit theorem certainly gives a good approximation in the center of the distribution of $S(t)$ around $ES(t)$, but it shows how dangerous it is to use the central limit theorem when it comes to considering probabilities

$$P(S(t) > y \mid N(t) = n(t)) = P\left(\frac{S(t) - n(t)\,EX_1}{\sqrt{n(t)\,\mathrm{var}(X_1)}} > \frac{y - n(t)\,EX_1}{\sqrt{n(t)\,\mathrm{var}(X_1)}} \right).$$

for large y. The normal approximation is poor if $x = (y - n(t)\,EX_1)/\sqrt{n(t)\,\mathrm{var}(X_1)}$ is too large. In particular, it can happen that the error bound on the right-hand side of (3.3.42) is larger than the approximated probability $1 - \Phi(x)$.

Example 3.3.13 (The tail of the distribution of $S(t)$ for subexponential claim sizes)
In this example we want to contrast the approximation of $P(S(t) > x)$ for $t \to \infty$ and fixed x, as provided by the central limit theorem, with an approximation for fixed t and large x. We assume the Cramér-Lundberg model and consider subexponential claim sizes. Therefore recall from p. 109 the definition of a subexponential distribution: writing $S_0 = 0$ and $S_n = X_1 + \cdots + X_n$ for the partial sums and $M_n = \max(X_1, \ldots, X_n)$ for the partial maxima of the iid claim size sequence (X_n), the distribution of X_1 and its distribution function F_{X_1} are said to be subexponential if

For every $n \ge 2$: $P(S_n > x) = P(M_n > x)\,(1 + o(1)) = n\,\overline{F}_{X_1}(x)(1 + o(1)),$

as $x \to \infty$. We will show that a similar relation holds if the partial sums S_n are replaced by the random sums $S(t)$.

We have, by conditioning on $N(t)$,

$$\frac{P(S(t) > x)}{\overline{F}_{X_1}(x)} = \sum_{n=0}^{\infty} P(N(t) = n) \frac{P(S_n > x)}{\overline{F}_{X_1}(x)} = \sum_{n=0}^{\infty} e^{-\lambda t} \frac{(\lambda t)^n}{n!} \frac{P(S_n > x)}{\overline{F}_{X_1}(x)}.$$

If we interchange the limit as $x \to \infty$ and the infinite series on the right-hand side, the subexponential property of F_{X_1} yields

$$\lim_{x \to \infty} \frac{P(S(t) > x)}{\overline{F}_{X_1}(x)} = \sum_{n=0}^{\infty} e^{-\lambda t} \frac{(\lambda t)^n}{n!} \lim_{x \to \infty} \frac{P(S_n > x)}{\overline{F}_{X_1}(x)}$$

$$= \sum_{n=0}^{\infty} e^{-\lambda t} \frac{(\lambda t)^n}{n!} n = EN(t) = \lambda t.$$

This is the analog of the subexponential property for the random sum $S(t)$. It shows that the central limit theorem is not a good guide in the tail of the distribution of $S(t)$; in this part of the distribution the heavy right tail of the claim size distribution determines the decay which is much slower than for the tail $\overline{\Phi}$ of the standard normal distribution.

We still have to justify the interchange of the limit as $x \to \infty$ and the infinite series $\sum_{n=0}^{\infty}$. We apply a domination argument. Namely, if we can find a sequence (f_n) such that

$$\sum_{n=0}^{\infty} \frac{(\lambda t)^n}{n!} f_n < \infty \quad \text{and} \quad \frac{P(S_n > x)}{\overline{F}_{X_1}(x)} \le f_n \quad \text{for all } x > 0, \quad (3.3.43)$$

then we are allowed to interchange these limits by virtue of the Lebesgue dominated convergence theorem; see Williams [78]. Recall from Lemma 3.2.24(3) that for any $\varepsilon > 0$ we can find a constant K such that

$$\frac{P(S_n > x)}{\overline{F}_{X_1}(x)} \le K (1 + \varepsilon)^n, \quad \text{for all } n \ge 1.$$

With the choice $f_n = K (1 + \varepsilon)^n$ for any $\varepsilon > 0$, it is not difficult to see that (3.3.43) is satisfied. □

Comments

The aim of this section was to show that an unsophisticated use of the normal approximation to the distribution of the total claim amount should be avoided, typically when one is interested in the probability of rare events, for example of $\{S(t) > x\}$ for x exceeding the expected claim amount $ES(t)$. In this case, other tools (asymptotic expansions for the distribution of $S(t)$, large deviation probabilities for the very large values x, saddlepoint approximations) can be used as alternatives. We refer to the literature mentioned in the text and to Embrechts et al. [29], Chapter 2, to get an impression of the complexity of the problem.

3.3.5 Approximation to the Distribution of the Total Claim Amount by Monte Carlo Techniques

One way out of the situation we encountered in Section 3.3.4 is to use the power and memory of modern computers to approximate the distribution of

$S(t)$. For example, if we knew the distributions of the claim number $N(t)$ and of the claim sizes X_i, we could simulate an iid sample $N_1, \ldots, N_m$ from the distribution of $N(t)$. Then we could draw iid samples

$$X_1^{(1)}, \ldots, X_{N_1}^{(1)}, \ldots, X_1^{(m)}, \ldots, X_{N_m}^{(m)}$$

from the distribution of X_1 and calculate iid copies of $S(t)$:

$$S_1 = \sum_{i=1}^{N_1} X_i^{(1)}, \ldots, S_m = \sum_{i=1}^{N_m} X_i^{(m)}.$$

The probability $P(S(t) \in A)$ for some Borel set A could be approximated by virtue of the strong law of large numbers:

$$\widehat{p}_m = \frac{1}{m} \sum_{i=1}^{m} I_A(S_i) \overset{a.s.}{\to} P(S(t) \in A) = p = 1 - q \quad \text{as } m \to \infty.$$

Notice that $m\,\widehat{p}_m \sim \text{Bin}(m, p)$. The approximation of p by the relative frequencies $\widehat{p}_m$ of the event A is called (crude) *Monte Carlo simulation*.

The rate of approximation could be judged by applying the central limit theorem with Berry-Esséen specification, see (3.3.41):

$$\sup_x (1 + |x|^3) \left| P\left(\frac{\widehat{p}_m - p}{\sqrt{pq/m}} \leq x \right) - \Phi(x) \right| \leq c \frac{p^3 q + (1-p)^3 p}{(\sqrt{pq})^3 \sqrt{m}} = c \frac{p^2 + q^2}{\sqrt{mpq}}.$$

$$(3.3.44)$$

We mentioned in the previous section that this bound is quite precise for a binomial distribution, i.e., for sums of Bernoulli random variables. This is encouraging, but for small probabilities p the Monte Carlo method is problematic. For example, suppose you want to approximate the probability $p = 10^{-k}$ for some $k \geq 1$. Then the rate on the right-hand side is of the order $10^{k/2}/\sqrt{m}$. This means you would need sample sizes m much larger than 10^k in order to make the right-hand side smaller than 1, and if one is interested in approximating small values of $\Phi(x)$ or $1 - \Phi(x)$, the sample sizes have to be chosen even larger. This is particularly unpleasant if one needs the whole distribution function of $S(t)$, i.e., if one has to calculate many probabilities of type $P(S(t) \leq y)$.

If one needs to approximate probabilities of very small order, say $p = 10^{-k}$ for some $k \geq 1$, then the crude Monte Carlo method does not work. This can be seen from the following argument based on the central limit theorem (3.3.44). The value p falls with 95% probability into the asymptotic confidence interval given by

$$\left[\widehat{p}_m - 1.96 \sqrt{pq/m} \; ; \; \widehat{p}_m + 1.96 \sqrt{pq/m} \right].$$

For practical purposes one would have to replace p in the latter relation by its estimator $\widehat{p}_m$. For small p this bound is inaccurate even if m is relatively large. One essentially has to compare the orders of magnitude of p and $1.96\sqrt{pq/m}$:

$$\frac{1.96\sqrt{pq/m}}{p} = \frac{1.96\sqrt{q}}{\sqrt{mp}} \approx 10^{k/2}\frac{1.96}{\sqrt{m}}.$$

This means we need sample sizes m much larger than 10^k in order to get a satisfactory approximation for p.

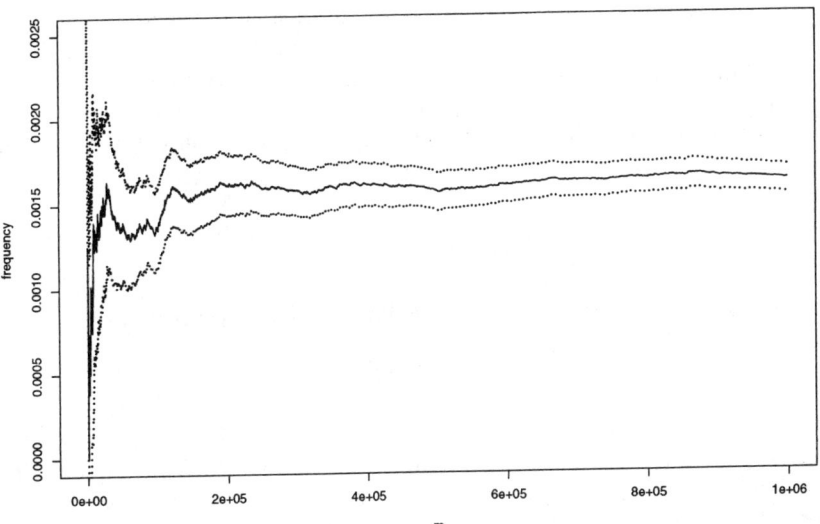

Figure 3.3.14 *Crude Monte Carlo simulation for the probability* $p = P(S(t) > ES(t) + 3.5\sqrt{\mathrm{var}(S(t))})$, *where* $S(t)$ *is the total claim amount in the Cramér-Lundberg model with Poisson intensity* $\lambda = 0.5$ *and Pareto distributed claim sizes with tail parameter* $\alpha = 3$, *scaled to variance 1. We have chosen* $t = 360$ *corresponding to one year. The intensity* $\lambda = 0.5$ *corresponds to expected inter-arrival times of 2 days. We plot* $\widehat{p}_m$ *for* $m \leq 10^6$ *and indicate 95% asymptotic confidence intervals prescribed by the central limit theorem. For* $m = 10^6$ *one has 1 618 values of* $S(t)$ *exceeding the threshold* $ES(t) + 3.5\sqrt{\mathrm{var}(S(t))}$, *corresponding to* $\widehat{p}_m = 0.001618$. *For* $m \leq 20\ 000$ *the estimates* $\widehat{p}_m$ *are extremely unreliable and the confidence bands are often wider than the approximated probability.*

The crude Monte Carlo approximation can be significantly improved for small probabilities p and moderate sample sizes m. Over the last 30 years special techniques such as *importance sampling* have been developed and run under the name of *rare event simulation*; see Asmussen [3, 4]. In an insurance context, rare events such as the WTC disaster or windstorm claims can have substantial impact on the insurance business; see Table 3.2.18. Therefore it is important to know that there are various techniques available which allow one to approximate such probabilities efficiently. By virtue of Poisson's limit theorem, rare events are more naturally approximated by Poisson probabilities. Approximations to the binomial distribution with small success probability by the Poisson distribution have been studied for a long time and optimal rates of this approximation were derived; see for example Barbour et al. [8]. Alternatively, the Poisson approximation is an important tool for rare events in the context of catastrophic or extremal events; see Embrechts et al. [29].

In the rest of this section we consider a statistical simulation technique which has become quite popular among statisticians and users of statistics over the last 20 years: Efron's [26] *bootstrap*. In contrast to the approximation techniques considered so far it does a priori not require any information about the distribution of the X_i's; all it uses is the information contained in the data available. In what follows, we focus on the case of an iid claim size sample $X_1, \ldots, X_n$ with common distribution function F and empirical distribution function

$$F_n(x) = \frac{1}{n} \sum_{i=1}^{n} I_{(-\infty, x]}(X_i), \quad x \in \mathbb{R}.$$

Then the Glivenko-Cantelli result (see Billingsley [13]) ensures that

$$\sup_x |F_n(x) - F(x)| \overset{\text{a.s.}}{\to} 0.$$

The latter relation has often been taken as a justification for replacing quantities depending on the *unknown* distribution function F by the same quantities depending on the *known* distribution function F_n. For example, in Section 3.2.3 we constructed the empirical mean excess function from the mean excess function in this way. The bootstrap extends this idea substantially: it suggests to *sample* from the empirical distribution function and to simulate pseudo-samples of iid random variables with distribution function F_n.

We explain the basic ideas of this approach. Let

$$x_1 = X_1(\omega), \ldots, x_n = X_n(\omega)$$

be the values of an observed iid sample which we consider as fixed in the sequel, i.e., the empirical distribution function F_n is a given discrete distribution function with equal probability at the x_i's. Suppose we want to approximate the distribution of a function $\theta_n = \theta_n(X_1, \ldots, X_n)$ of the data, for example of the sample mean

$$\overline{X}_n = \frac{1}{n} \sum_{i=1}^{n} X_i.$$

The bootstrap is then given by the following algorithm.

(a) Draw with replacement from the distribution function F_n the iid realizations

$$X_1^*(1), \ldots, X_n^*(1), \ldots, X_1^*(B), \ldots, X_n^*(B)$$

for some large number B. *In principle*, using computer power we could make B *arbitrarily large*.

(b) Calculate the iid sample

$$\theta_n^*(1) = \theta_n(X_1^*(1), \ldots, X_n^*(1)), \ldots, \theta_n^*(B) = \theta_n(X_1^*(B), \ldots, X_n^*(B)).$$

In what follows we write $X_i^* = X_i^*(1)$ and $\theta_n^* = \theta_n^*(1)$.

(c) Approximate the distribution of θ_n^* and its characteristics such as moments, quantiles, etc., either by direct calculation or by using the strong law of large numbers.

We illustrate the meaning of this algorithm for the sample mean.

Example 3.3.15 (The bootstrap sample mean)
The sample mean $\theta_n = \overline{X}_n$ is an unbiased estimator of the expectation $\theta = EX_1$, provided the latter expectation exists and is finite. The *bootstrap sample mean* is the quantity

$$\overline{X}_n^* = \frac{1}{n} \sum_{i=1}^n X_i^*.$$

Since the (conditionally) iid X_i^*'s have the discrete distribution function F_n,

$$E^*(X_1^*) = E_{F_n}(X_1^*) = \frac{1}{n} \sum_{i=1}^n x_i = \overline{x}_n,$$

$$\mathrm{var}^*(X_1^*) = \mathrm{var}_{F_n}(X_1^*) = \frac{1}{n} \sum_{i=1}^n (x_i - \overline{x}_n)^2 = s_n^2.$$

Now using the (conditional) independence of the X_i^*'s, we obtain

$$E^*(\overline{X}_n^*) = \frac{1}{n} \sum_{i=1}^n E^*(X_i^*) = E^*(X_1^*) = \overline{x}_n,$$

$$\mathrm{var}^*(\overline{X}_n^*) = \frac{1}{n^2} \sum_{i=1}^n \mathrm{var}^*(X_i^*) = n^{-1}\, \mathrm{var}^*(X_1^*) = n^{-1} s_n^2.$$

For more complicated functionals of the data it is in general not possible to get such simple expressions as for $\overline{X}_n^*$. For example, suppose you want to calculate the distribution function of $\overline{X}_n^*$ at x:

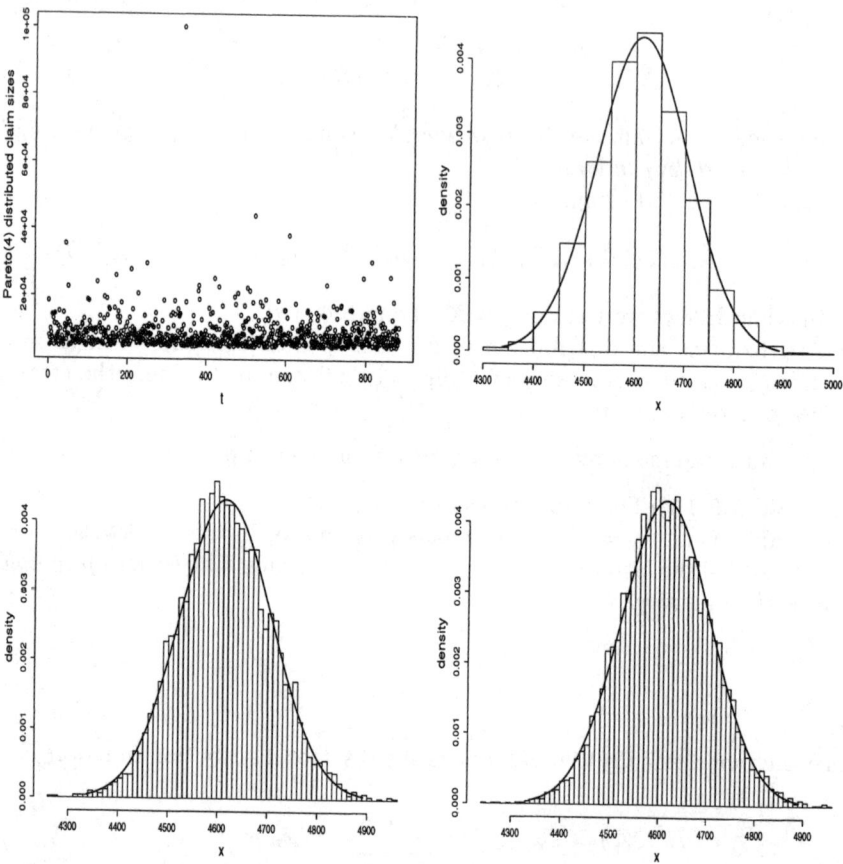

Figure 3.3.16 *The bootstrap for the sample mean of* 3 000 *Pareto distributed claim sizes with tail index* $\alpha = 4$; *see Table 3.2.19. The largest value is* 10 000 *$US. The claim sizes* X_n *which exceed the threshold of* 5 000 *$US are shown in the* top left *graph. The* top right, bottom left, bottom right *graphs show histograms of the bootstrap sample mean with bootstrap sample size* $B = 2$ 000 *(left),* $B = 5$ 000 *(middle) and* $B = 10$ 000 *(right), respectively. For comparison we draw the normal density curve with the mean and variance of the data in the histograms.*

$$P^*(\overline{X}_n^* \le x) = E_{F_n}\left(I_{(-\infty,x]}(\overline{X}_n^*)\right) = \frac{1}{n^n} \sum_{i_1=1}^{n} \cdots \sum_{i_n=1}^{n} I_{(-\infty,x]}\left(\frac{1}{n}\sum_{j=1}^{n} x_{i_j}\right).$$

This means that, in principle, one would have to evaluate n^n terms and sum them up. Even with modern computers and for small sample sizes such as $n = 10$ this would be a too difficult computational problem. On the other hand, the Glivenko-Cantelli result allows one to approximate $P^*(\overline{X}_n^* \le x)$

arbitrarily closely by choosing a large bootstrap sample size B:

$$\sup_x \left| \frac{1}{B} \sum_{i=1}^{B} I_{(-\infty,x]}(\overline{X}_n^*(i)) - P^*(\overline{X}_n^* \le x) \right| \to 0 \quad \text{as } B \to \infty,$$

with probability 1, where this probability refers to a probability measure which is constructed from F_n. In practical simulations one can make B very large. Therefore it is in general not considered a problem to approximate the distribution of functionals of $X_1^*, \ldots, X_n^*$ as accurately as one wishes.   □

The bootstrap is mostly used to approximate the distributional characteristics of functionals θ_n of the data such as the expectation, the variance and quantiles of θ_n in a rather unsophisticated way. In an insurance context, the method allows one to approximate the distribution of the aggregated claim sizes $n\overline{X}_n = X_1 + \cdots + X_n$ by its bootstrap version $X_1^* + \cdots + X_n^*$ or of the total claim amount $S(t)$ conditionally on the claim number $N(t)$ by approximation through the bootstrap version $X_1^* + \cdots + X_{N(t)}^*$, and bootstrap methods can be applied to calculate confidence bands for the parameters of the claim number and claim size distributions.

    Thus it seems as if the bootstrap solves all statistical problems of this world without too much sophistication. This was certainly the purpose of its inventor Efron [26], see also the text by Efron and Tibshirani [27]. However, the replacement of the X_i's with distribution function F with the corresponding bootstrap quantities X_i^* with distribution function F_n in a functional $\theta_n(X_1, \ldots, X_n)$ has actually a continuity problem. This replacement does not always work even for rather simple functionals of the data; see Bickel and Freedman [11] for some counterexamples. Therefore one has to be careful; as for the crude Monte Carlo method considered above the *naive* bootstrap can one lead into the wrong direction, i.e., the bootstrap versions θ_n^* can have distributions which are far away from the distribution of θ_n. Moreover, in order to show that the bootstrap approximation "works", i.e., it is close to the distribution of θ_n, one needs to apply asymptotic techniques for $n \to \infty$. This is slightly disappointing because the original idea of the bootstrap was to be applicable to small sample size.

    As a warning we also mention that the naive bootstrap for the total claim amount does not work if one uses very heavy-tailed distributions. Then bootstrap sampling forces one to draw the largest values in the sample too often, which leads to deviations of the bootstrap distribution from the distribution of θ_n; see Figure 3.3.17 for an illustration of this phenomenon. Moreover, the bootstrap does not solve the problem of calculating the probability of rare events such as $P(S(t) > x)$ for values x far beyond the mean $ES(t)$; see the previous discussions. Since the empirical distribution function stops increasing at the maximum of the data, the bootstrap does not extrapolate into the tails of the distribution of the X_i's. For this purpose one has to depend on special parametric or semi-parametric methods such as those provided in extreme value theory; cf. Embrechts et al. [29], Chapter 6.

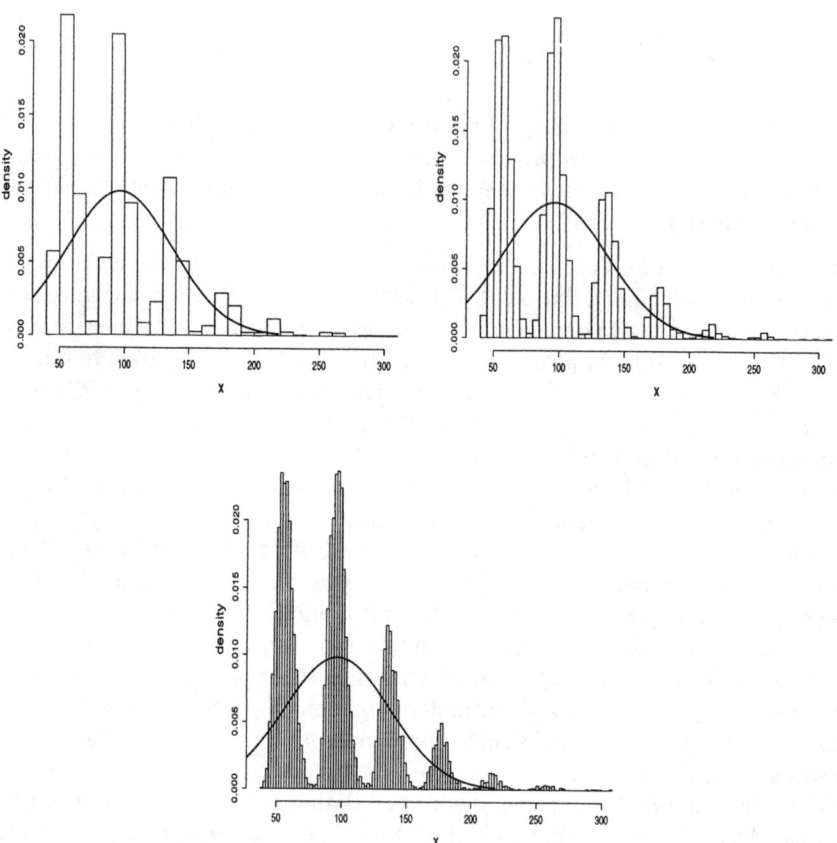

Figure 3.3.17 *The bootstrap for the sample mean of* 3 000 *Pareto distributed claim sizes with tail index* $\alpha = 1$. *The graphs show histograms of the bootstrap sample mean with bootstrap sample size* $B = 2\,000$ *(top left),* $B = 5\,000$ *(top right) and* $B = 10\,000$ *(bottom). For comparison we draw the normal density curve with the sample mean and sample variance of the data in the histograms. It is known that the Pareto distribution with tail index* $\alpha = 1$ *does not satisfy the central limit theorem with normal limit distribution (e.g. [29], Chapter 2), but with a skewed Cauchy limit distribution. Therefore the misfit of the normal distribution is not surprising, but the distribution of the bootstrap sample mean is also far from the Cauchy distribution which has a unimodal density. In the case of infinite variance claim size distributions,* the (naive) bootstrap does not work for the sample mean.

Comments

Monte Carlo simulations and the bootstrap are rather recent computer-based methods, which have an increasing appeal since the quality of the computers has enormously improved over the last 15-20 years. These methods provide an ad hoc approach to problems whose exact solution had been considered hopeless. Nevertheless, none of these methods is perfect. Pitfalls may occur even in rather simple cases. Therefore one should not use these methods without consulting the relevant literature. Often theoretical means such as the central limit theorem of Section 3.3.4 give the same or even better approximation results. Simulation should only be used if nothing else works.

The book by Efron and Tibshirani [27] is an accessible introduction to the bootstrap. Books such as Hall [41] or Mammen [56] show the limits of the method, but also require knowledge on mathematical statistics.

Asmussen's lecture notes [3] are a good introduction to the simulation of stochastic processes and distributions, see also Chapter X in Asmussen [4] and the references cited therein. That chapter is devoted to simulation methodology, in particular for rare events. Survey papers about rare event simulation include Asmussen and Rubinstein [7] and Heidelberger [42]. Rare event simulation is particularly difficult when heavy-tailed distributions are involved. This is, for example, documented in Asmussen et al. [6].

Exercises

Section 3.3.1

(1) Decomposition of the claim size space for discrete distribution.

(a) Let $N_1, \ldots, N_n$ be independent Poisson random variables with $N_i \sim \text{Pois}(\lambda_i)$ for some $\lambda_i > 0$, $x_1, \ldots, x_n$ be positive numbers. Show that $x_1 N_1 + \cdots + x_n N_n$ has a compound Poisson distribution.

(b) Let $S = \sum_{k=1}^N X_k$ be compound Poisson where $N \sim \text{Pois}(\lambda)$, independent of the iid claim size sequence (X_k) and $P(X_1 = x_i) = p_i$, $i = 1, \ldots, n$, for some distribution (p_i). Show that $S \stackrel{d}{=} x_1 N_1 + \cdots + x_n N_n$ for appropriate independent Poisson variables $N_1, \ldots, N_n$.

(c) Assume that the iid claim sizes X_k in an insurance portfolio have distribution $P(X_k = x_i) = p_i$, $i = 1, \ldots, n$. The sequence (X_k) is independent of the Poisson claim number N with parameter λ. Consider a disjoint partition $A_1, \ldots, A_m$ of the possible claim sizes $\{x_1, \ldots, x_n\}$. Show that the total claim amount $S = \sum_{k=1}^N X_k$ has the same distribution as

$$\sum_{i=1}^m \sum_{k=1}^{N_i} X_k^{(i)},$$

where $N_i \sim \text{Pois}(\lambda_i)$, $\lambda_i = \lambda \sum_{k \in A_i} p_k$, are independent Poisson variables, independent of $(X_k^{(i)})$ and for each i, $X_k^{(i)}$, $k = 1, 2, \ldots$, are iid with distribution $P(X_k^{(i)} = x_l) = p_l / \sum_{s \in A_i} p_s$. This means that one can split the claim sizes into distinct categories (for example one can introduce layers $A_i = (a_i, b_i]$ for the claim sizes or one can split the claims into small and large ones according as $x_i \leq u$ or $x_i > u$ for a threshold u) and consider the total claim amount from each category as a compound Poisson variable.

(2) Consider the total claim amount $S(t) = \sum_{i=1}^{N(t)} X_i$ in the Cramér-Lundberg model for fixed t, where N is homogeneous Poisson and independent of the claim size sequence (X_i).

(a) Show that

$$S(t) \overset{d}{=} \sum_{i=1}^{N_1(t)+N_2(t)} X_i \overset{d}{=} \sum_{i=1}^{N_1(t)} X_i + \sum_{i=1}^{N_2(t)} X_i',$$

where N_1 and N_2 are independent homogeneous Poisson processes with intensities λ_1 and λ_2, respectively, such that $\lambda_1 + \lambda_2 = \lambda$, (X_i') is an independent copy of (X_i), and N_1, N_2, (X_i) and (X_i') are independent.

(b) Show relation (3.3.29) by calculating the joint characteristic functions of the left- and right-hand expressions.

(3) We consider the mixed Poisson processes $N_i(t) = \tilde{N}_i(\theta_i t)$, $t \geq 0$, $i = 1, \ldots, n$. Here $\tilde{N}_i$ are mutually independent standard homogeneous Poisson processes, θ_i are mutually independent positive mixing variables, and $(\tilde{N}_i)$ and (θ_i) are independent. Consider the independent *compound mixed Poisson sums*

$$S_j = \sum_{i=1}^{N_j(1)} X_i^{(j)}, \quad j = 1, \ldots, n,$$

where $(X_i^{(j)})$ are iid copies of a sequence (X_i) of iid positive claim sizes, independent of (N_j). Show that $S = S_1 + \cdots + S_n$ is again a compound mixed Poisson sum with representation

$$S \overset{d}{=} \sum_{i=1}^{\tilde{N}_1(\theta_1+\cdots+\theta_n)} X_i.$$

(4) Let $S = \sum_{i=1}^{N} X_i$ be the total claim amount at a fixed time t, where the claim number N and the iid claim size sequence (X_i) are independent.

(a) Show that the *Laplace-Stieltjes transform* of S, i.e., $\hat{f}_S(s) = m_S(-s) = Ee^{-sS}$ always exists for $s \geq 0$.

(b) Show that

$$P(S > x) \leq ce^{-hx} \quad \text{for all } x > 0, \text{ some } c > 0, \tag{3.3.45}$$

if $m_S(h) < \infty$ for some $h > 0$. Show that (3.3.45) implies that the moment generating function $m_S(s) = Ee^{sS}$ is finite in some neighborhood of the origin.

(5) Recall the negative binomial distribution

$$p_k = \binom{v+k-1}{k} p^v (1-p)^k, \quad k = 0, 1, 2, \ldots, \quad p \in (0, 1), \quad v > 0.$$

$$\tag{3.3.46}$$

Recall from Example 2.3.3 that the negative binomial process $(N(t))_{t \geq 0}$ is a mixed standard homogeneous Poisson process with mixing variable θ with gamma $\Gamma(\gamma, \beta)$ density

$$f_\theta(x) = \frac{\beta^\gamma}{\Gamma(\gamma)} x^{\gamma-1} e^{-\beta x}, \quad x > 0.$$

Choosing $v = \gamma$ and $p = \beta/(1+\beta)$, $N(1)$ then has distribution (3.3.46).

(a) Use this fact to calculate the characteristic function of a negative binomial random variable with parameters p and ν.

(b) Let $N \sim \text{Pois}(\lambda)$ be the number of accidents in a car insurance portfolio in a given period, X_i the claim size in the ith accident and assume that the claim sizes X_i are iid positive and integer-valued with distribution

$$P(X_i = k) = \frac{k^{-1} p^k}{-\log(1-p)}, \quad k = 1, 2, \ldots.$$

for some $p \in (0,1)$. Verify that these probabilities define a distribution, the so-called *logarithmic distribution*. Calculate the characteristic function of the compound Poisson variable $S = \sum_{i=1}^{N} X_i$. Verify that it has a negative binomial distribution with parameters $\tilde{v} = -\lambda/\log(1-p)$ and $\tilde{p} = 1-p$. Hence a random variable with a negative binomial distribution has representation as a compound Poisson sum with logarithmic claim size distribution.

(6) A distribution F is said to be *infinitely divisible* if for every $n \geq 1$, its characteristic function ϕ can be written as a product of characteristic functions ϕ_n:

$$\phi(s) = (\phi_n(s))^n, \quad s \in \mathbb{R}.$$

In other words, for every $n \geq 1$, there exist iid random variables $Y_{n,1}, \ldots, Y_{n,n}$ with common characteristic function ϕ_n such that for a random variable Y with distribution F the following identity in distribution holds:

$$Y \stackrel{d}{=} Y_{n,1} + \cdots + Y_{n,n}.$$

Almost every familiar distribution with unbounded support which is used in statistics or probability theory has this property although it is often very difficult to prove this fact for concrete distributions. We refer to Lukacs [54] or Sato [71] for more information on this class of distributions.

(a) Show that the normal, Poisson and gamma distributions are infinitely divisible.

(b) Show that the distribution of a compound Poisson variable is infinitely divisible.

(c) Consider a compound Poisson process $S(t) = \sum_{i=1}^{N(t)} X_i$, $t \geq 0$, where N is a homogeneous Poisson process on $[0, \infty)$ with intensity $\lambda > 0$, independent of the iid claim sizes X_i. Show that the process S obeys the following infinite divisibility property: for every $n \geq 1$ there exist iid compound Poisson processes S_i such that $S \stackrel{d}{=} S_1 + \cdots + S_n$, where $\stackrel{d}{=}$ refers to identity of the finite-dimensional distributions. Hint: Use the fact that S and S_i have independent and stationary increments.

Section 3.3.3

(7) The (a, b)-class of distributions.

(a) Verify the (a, b)-condition

$$q_n = P(N = n) = \left(a + \frac{b}{n}\right) q_{n-1} \tag{3.3.47}$$

for the Poisson, binomial and negative binomial claim number distributions (q_n) and appropriate choices of the parameters a, b. Determine the region R of possible (a, b)-values for these distributions.

(b) Show that the (a, b)-condition (3.3.47) for values $(a, b) \notin R$ does not define a probability distribution (q_n) of a random variable N with values in $\mathbb{N}_0$.

(c) Show that the Poisson, binomial and negative binomial distributions are the only possible distributions on $\mathbb{N}_0$ satisfying an (a, b)-condition, i.e., (3.3.47) implies that (q_n) is necessarily Poisson, binomial or negative binomial, depending on the choice of $(a, b) \in R$.

Sections 3.3.4 and 3.3.5

(8) Consider an iid sample $X_1, \ldots, X_n$ and the corresponding empirical distribution function:

$$F_n(x) = \frac{1}{n} \#\{i \leq n : X_i \leq x\}.$$

By X^* we denote any random variable with distribution function F_n, given $X_1, \ldots, X_n$.

(a) Calculate the expectation, the variance and the third absolute moment of X^*.

(b) For (conditionally) iid random variables X_i^*, $i = 1, \ldots, n$, with distribution function F_n calculate the mean and variance of the sample mean $\overline{X}_n^* = n^{-1} \sum_{i=1}^{n} X_i^*$.

(c) Apply the strong law of large numbers to show that the limits of $E^*(\overline{X}_n^*)$ and $n\mathrm{var}^*(\overline{X}_n^*)$ as $n \to \infty$ exist and coincide with their deterministic counterparts EX_1 and $\mathrm{var}(X_1)$, provided the latter quantities are finite. Here E^* and var^* refer to expectation and variance with respect to the distribution function F_n of the (conditionally) iid random variables X_i^*'s.

(d) Apply the Berry-Esséen inequality to

$$P^* \left(\frac{\sqrt{n}}{\sqrt{\mathrm{var}^*(X_1^*)}} (\overline{X}_n^* - E^*(\overline{X}_n^*)) \leq x \right) - \Phi(x)$$

$$= P \left(\frac{\sqrt{n}}{\sqrt{\mathrm{var}^*(X_1^*)}} (\overline{X}_n^* - E^*(\overline{X}_n^*)) \leq x \,\Big|\, X_1, \ldots, X_n \right) - \Phi(x),$$

where Φ is the standard normal distribution function and show that the (conditional) central limit theorem applies[9] to (X_i^*) if $E|X_1|^3 < \infty$, i.e., the above differences converge to 0 with probability 1.

Hint: It is convenient to use the elementary inequality

$$|x + y|^3 \leq (2 \max(|x|, |y|))^3 = 8 \max(|x|^3, |y|^3) \leq 8 (|x|^3 + |y|^3), \quad x, y \in \mathbb{R}.$$

(9) Let $X_1, X_2, \ldots$ be an iid sequence with finite variance (without loss of generality assume $\mathrm{var}(X_1) = 1$) and mean zero. Then the central limit theorem and the continuous mapping theorem (see Billingsley [12]) yield

$$T_n = n (\overline{X}_n)^2 = \left(\frac{1}{\sqrt{n}} \sum_{i=1}^{n} X_i \right)^2 \xrightarrow{d} Y^2,$$

where Y has a standard normal distribution. The naive bootstrap version of T_n is given by

---

[9] As a matter of fact, the central limit theorem applies to (X_i^*) under the weaker assumption $\mathrm{var}(X_1) < \infty$; see Bickel and Freedman [11].

$$T_n^* = n\,(\overline{X}_n^*)^2 = \left(\frac{1}{\sqrt{n}}\sum_{i=1}^n X_i^*\right)^2 ,$$

where (X_i^*) is an iid sequence with common empirical distribution function F_n based on the sample $X_1,\ldots,X_n$, i.e., (X_i^*) are iid, conditionally on $X_1,\ldots,X_n$.

(a) Verify that the bootstrap does not work for T_n^* by showing that (T_n^*) has no limit distribution with probability 1. In particular, show that the following limit relation does not hold as $n \to \infty$:

$$P^*(T_n^* \le x) = P(T_n^* \le x \mid X_1,\ldots,X_n)$$
$$\to P(Y^2 \le x), \quad x \ge 0. \tag{3.3.48}$$

Hints: (i) You may assume that we know that the central limit theorem

$$P^*(\sqrt{n}(\overline{X}_n^* - \overline{X}_n) \le x) \to \Phi(x) \quad \text{a.s.}, \quad x \in \mathbb{R},$$

holds as $n \to \infty$; see Exercise 8 above.
(ii) Show that $(\sqrt{n}\,\overline{X}_n)$ does not converge with probability 1.

(b) Choose an appropriate centering sequence for (T_n^*) and propose a modified bootstrap version of T_n^* which obeys the relation (3.3.48).

(10) Let (X_i^*) be a (conditionally) iid bootstrap sequence corresponding to the iid sample $X_1,\ldots,X_n$.

(a) Show that the bootstrap sample mean $\overline{X}_n^*$ has representation

$$\overline{X}_n^* \stackrel{d}{=} \frac{1}{n}\sum_{j=1}^n X_j \sum_{i=1}^n I_{((j-1)/n,\,j/n]}(U_i),$$

where (U_i) is an iid $U(0,1)$ sequence, independent of (X_i).

(b) Write

$$M_{n,j} = \sum_{i=1}^n I_{((j-1)/n,\,j/n]}(U_i).$$

Show that the vector $(M_{n,1},\ldots,M_{n,n})$ has a multinomial $\mathrm{Mult}(n; n^{-1},\ldots,n^{-1})$ distribution.

3.4 Reinsurance Treaties

In this section we introduce some reinsurance treaties which are standard in the literature. For the sake of illustration we assume the Cramér-Lundberg model with iid positive claim sizes X_i and Poisson intensity $\lambda > 0$.

Reinsurance treaties are mutual agreements between different insurance companies with the aim to reduce the risk in a particular insurance portfolio by sharing the risk of the occurring claims as well as the premium in this portfolio. In a sense, reinsurance is insurance for insurance companies. Reinsurance is a necessity for portfolios which are subject to catastrophic risks such as earthquakes, failure of nuclear power stations, major windstorms, industrial fire, tanker accidents, flooding, war, riots, etc. Often various insurance companies have mutual agreements about reinsuring certain parts of their portfolios. Major insurance companies such a Swiss and Munich Re or Lloyd's have specialized in reinsurance products and belong to the world's largest companies of their kind.

It is convenient to distinguish between two different types of reinsurance treaties:

- treaties of *random walk type*,
- treaties of *extreme value type*.

These names refer to the way how the treaties are constructed: *either* the total claim amount $S(t)$ (or a modified version of it) *or* some of the largest order statistics of the claim size sample are used for the construction of the treaty.
We start with *reinsurance treaties of random walk type*.

(1) *Proportional reinsurance.* This is a common form of reinsurance for claims of "moderate" size. Here simply a fraction $p \in (0,1)$ of each claim (hence the pth fraction of the whole portfolio) is covered by the reinsurer. Thus the reinsurer pays for the amount $R_{\mathrm{Prop}}(t) = p\,S(t)$ whatever the size of the claims.

(2) *Stop-loss reinsurance.* The reinsurer covers losses in the portfolio exceeding a well-defined limit K, the so-called *ceding company's retention level.* This means that the reinsurer pays for $R_{\mathrm{SL}}(t) = (S(t) - K)_+$, where $x_+ = \max(x, 0)$. This type of reinsurance is useful for protecting the company against insolvency due to excessive claims on the coverage.[10]

(3) *Excess-of-loss reinsurance.* The reinsurance company pays for all individual losses in excess of some limit D, i.e., it covers $R_{\mathrm{ExL}}(t) = \sum_{i=1}^{N(t)}(X_i - D)_+$. The limit D has various names in the different branches of insurance. In life insurance, it is called the *ceding company's retention level.* In non-life insurance, where the size of loss is unknown in advance, D is called *deductible.* The reinsurer may in reality not insure the whole risk exceeding some limit D but rather buy a layer of reinsurance corresponding to coverage of claims in the interval $(D_1, D_2]$. This can be done directly or by itself obtaining reinsurance from another reinsurer.

Notice that any of the quantities $R_i(t)$ defined above is closely related to the total claim amount $S(t)$; the same results and techniques which were developed in the previous sections can be used to evaluate the distribution and the distributional characteristics of $R_i(t)$. For example,

---

[10] The stop-loss treaty bears some resemblance with the terminal value of a so-called *European call option.* In this context, $S(t)$ is the price of a risky asset at time t such (as a share price, a foreign exchange rate or a stock index) and $(S(T) - K)+$ is the value of the option with strike price K at time T of maturity. Mathematical finance deals with the pricing and hedging of such contracts; we refer to Björk [15] for a mathematical introduction to the field and to Mikosch [57] for an elementary approach.

$$P(R_{\mathrm{SL}}(t) \le x) = P(S(t) \le K) + P(K < S(t) \le x + K), \quad x \ge 0,$$

and the processes R_{Prop} and R_{ExL} have total claim amount structure with claim sizes $p\,X_i$ and $(X_i - D)_+$, respectively.

Treaties of *extreme value type* aim at covering the largest claims in a portfolio. Consider the iid claim sizes $X_1, \ldots, X_{N(t)}$ which occurred up to time t and the corresponding ordered sample

$$X_{(1)} \le \cdots \le X_{(N(t))}.$$

(4) *Largest claims reinsurance.* At the time when the contract is underwritten (i.e., at $t = 0$) the reinsurance company guarantees that the k largest claims in the time frame $[0, t]$ will be covered. For example, the company will cover the 10 largest annual claims in a portfolio over a period of 5 years, say.

This means that one has to study the quantity

$$R_{\mathrm{LC}}(t) = \sum_{i=1}^{k} X_{(N(t)-i+1)}$$

either for a fixed k or for a k which grows sufficiently slowly with t.

(5) *ECOMOR reinsurance (Excédent du coût moyen relatif).* This form of a treaty can be considered as an excess-of-loss reinsurance with a random deductible which is determined by the kth largest claim in the portfolio. This means that the reinsurer covers the claim amount

$$R_{\mathrm{ECOMOR}}(t) = \sum_{i=1}^{N(t)} \left(X_{(N(t)-i+1)} - X_{(N(t)-k+1)} \right)_+$$

$$= \sum_{i=1}^{k-1} X_{(N(t)-i+1)} - (k-1) X_{(N(t)-k+1)}$$

for a fixed number $k \ge 2$.

Treaties of random walk type can be studied by using tools for random walks such as the strong law of large numbers, the central limit theorem and ruin probabilities as considered in Chapter 4. In contrast to the latter, treaties of extreme value type need to be studied by extreme value theory techniques which are beyond the scope of this course. We refer to Embrechts et al. [29] for an introduction, in particular, to Section 8.7, where reinsurance treaties are considered.

With the mathematical theory we have learned so far we can solve some problems which are related to reinsurance treaties:

(1) How many claim sizes can occur in a layer $(D_1, D_2]$ or (D_1, ∞) up to time t?

(2) What can we say about the distribution of the largest claims?

It turns out that we can use similar techniques for answering these questions: we embed the pairs (T_i, X_i) in a Poisson process.

We start with the first question.

Example 3.4.1 (Distribution of the number of claim sizes in a layer)
We learned in Section 2.1.8 that (T_i, X_i) constitute the points of a Poisson process M with state space $[0, \infty)^2$ and mean measure $(\lambda \operatorname{Leb}) \times F_{X_1}$, where Leb is Lebesgue measure on $[0, \infty)$. Concerning question (1), we are interested in the distribution of the quantity

$$M((0, t] \times A) = \#\{i \geq 1 : X_i \in A, T_i \leq t\} = \sum_{i=1}^{N(t)} I_A(X_i)$$

for some Borel set A and fixed $t > 0$. Since M is a Poisson process with mean measure $(\lambda \operatorname{Leb}) \times F_{X_1}$, we immediately have the distribution of $M((0, t] \times A)$:

$$M((0, t] \times A) \sim \operatorname{Pois}(F_{X_1}(A) \lambda t).$$

This solves problem (1) for limited layers $A_1 = (D_1, D_2]$ or unlimited layers $A_2 = (D_2, \infty]$. From the properties of the Poisson process M we also know that $M((0, t] \times A_1)$ and $M((0, t] \times A_2)$ are independent. Even more is true: we know from Section 3.3.2 that the corresponding total claim amounts $\sum_{i=1}^{N(t)} X_i I_{A_1}(X_i)$ and $\sum_{i=1}^{N(t)} X_i I_{A_2}(X_i)$ are independent. □

As regards the second question, we can give exact formulae for the distribution of the largest claims:

Example 3.4.2 (Distribution of the largest claim sizes)
We proceed in a similar way as in Example 3.4.1 and use the same notation. Observe that

$$\{X_{(N(t)-k+1)} \leq x\} = \{M((0, t] \times (x, \infty)) < k\}.$$

Since $M((0, t] \times (x, \infty)) \sim \operatorname{Pois}(\overline{F}_{X_1}(x) \lambda t)$,

$$P(X_{(N(t)-k+1)} \leq x) = \sum_{i=0}^{k-1} e^{-\overline{F}_{X_1}(x) \lambda t} \frac{(\overline{F}_{X_1}(x) \lambda t)^i}{i!}.$$

□

As a matter of fact, it is much more complicated to deal with sums of order statistics as prescribed by the largest claims and the ECOMOR treaties. In general, it is impossible to give exact distributional characteristics of R_{LC} and R_{ECOMOR}. One of the few exceptions is the case of exponential claim sizes.

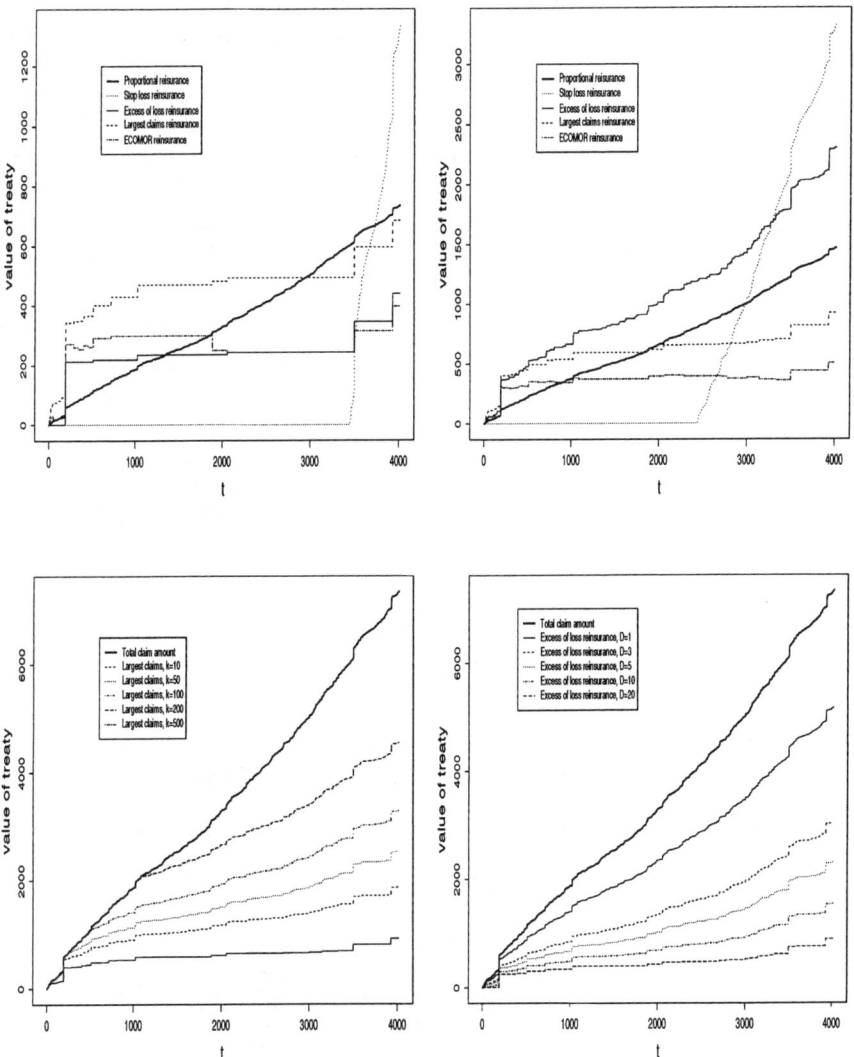

Figure 3.4.3 *The values of the reinsurance treaties as a function of time for the Danish fire insurance data from January 1, 1980, until 31 December, 1990; see Section 2.1.7 for a description of the data. Prices on the y-axis are in thousands of Kroner. Top left: Proportional with p = 0.1, stop-loss with K = 6 millions, excess-of-loss with D = 50 000, largest claims and ECOMOR with k = 5. Top right: Proportional with p = 0.2, stop-loss with K = 4 millions, excess-of-loss with D = 5 000, largest claims and ECOMOR with k = 10. Notice the differences in scale on the y-axis. Bottom left: Largest claims reinsurance for different claim numbers k. Bottom right: Excess-of-loss reinsurance for different deductibles D.*

Example 3.4.4 (Treaties of extreme value type for exponential claim sizes) Assume that the claim sizes are iid $\text{Exp}(\gamma)$ distributed. From Exercise 13 on p. 55 we learn that the order statistics of the sample $X_1, \ldots, X_n$ have the representation

$$\left(X_{(1)}, \ldots, X_{(n)}\right) \overset{d}{=} \left(\frac{X_n}{n}, \frac{X_n}{n} + \frac{X_{n-1}}{n-1}, \ldots, \frac{X_n}{n} + \frac{X_{n-1}}{n-1} + \cdots + \frac{X_2}{2}, \right.$$
$$\left. \frac{X_n}{n} + \frac{X_{n-1}}{n-1} + \cdots + \frac{X_1}{1}\right).$$

This implies that

$$\sum_{i=1}^{k} X_{(n-i+1)} \overset{d}{=} \sum_{i=1}^{k} \left(\frac{X_i}{i} + \cdots + \frac{X_n}{n}\right) = \sum_{i=1}^{k} X_i + k \sum_{i=k+1}^{n} \frac{X_i}{i}$$

and

$$\sum_{i=1}^{k-1} X_{(n-i+1)} - (k-1) X_{(n-k+1)} \overset{d}{=} X_1 + \cdots + X_{k-1}.$$

Hence the ECOMOR treaty has distribution

$$R_{\text{ECOMOR}}(t) \overset{d}{=} X_1 + \cdots + X_{k-1} \sim \Gamma(k-1, \gamma), \quad k \geq 2,$$

irrespective of t. The largest claims treaty has a less attractive distribution, but one can determine a limit distribution as $t \to \infty$. First observe that for every $t \geq 0$,

$$R_{\text{LC}}(t) \overset{d}{=} \sum_{i=1}^{k} X_i + k \sum_{i=k+1}^{N(t)} \frac{X_i}{i}$$
$$= \sum_{i=1}^{k} X_i + k \, EX_1 \sum_{i=k+1}^{N(t)} i^{-1} + k \sum_{i=k+1}^{N(t)} \frac{X_i - EX_1}{i}$$

The homogeneous Poisson process has the property $N(t) \overset{\text{a.s.}}{\to} \infty$ as $t \to \infty$ since it satisfies the strong law of large numbers $N(t)/t \overset{\text{a.s.}}{\to} \lambda$. Therefore,

$$\sum_{i=k+1}^{N(t)} \frac{X_i - EX_1}{i} \overset{\text{a.s.}}{\to} \sum_{i=k+1}^{\infty} \frac{X_i - EX_1}{i}.$$

The existence of the limit on the right-hand side is justified by Lemma 2.2.6 and the fact that the infinite series $\sum_{i=1}^{\infty} i^{-1}(X_i - EX_1)$ converges a.s. This statement can be verified by using the 3-series theorem or by observing that the infinite series has finite variance, cf. Billingsley [13], Theorems 22.6 and 22.8. It is well-known that $\sum_{i=k+1}^{n} i^{-1} = (1 + o(1)) \log n$ as $n \to \infty$. We conclude that, with probability 1, as $t \to \infty$,

$$\sum_{i=k+1}^{N(t)} i^{-1} = (1 + o(1)) \log N(t) = (1 + o(1)) \left[\log(N(t)/(\lambda t)) + \log(\lambda t)\right]$$

$$= (1 + o(1)) \log(\lambda t),$$

where we also used the strong law of large numbers for $N(t)$. Collecting the above limit relations, we end up with

$$R_{\mathrm{LC}}(t) - k\gamma^{-1} \log(\lambda t) \xrightarrow{d} \sum_{i=1}^{k} X_i + k \sum_{i=k+1}^{\infty} i^{-1}(X_i - \gamma^{-1}). \qquad (3.4.49)$$

The limiting distribution can be evaluated by using Monte Carlo methods; see Figure 3.4.5.   □

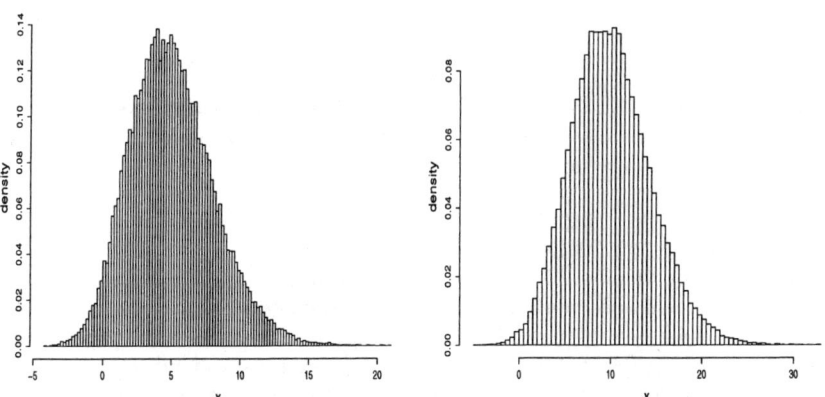

Figure 3.4.5 *Histogram of 50 000 iid realizations of the limiting distribution in (3.4.49) with $k = 5$ (left), $k = 10$ (right), and $\lambda = \gamma = 1$.*

Comments

Over the last few years, traditional reinsurance has been complemented by financial products which are sold by insurance companies. Those include catastrophe insurance bonds or derivatives such as options and futures based on some catastrophe insurance index comparable to a composite stock index such as the S&P 500, the Dow Jones, DAX, etc. This means that reinsurance has attracted the interest of a far greater audience. The interested reader is referred to Section 8.7 in Embrechts et al. [29] and the references therein for an introduction to this topic. The websites of Munich Re WWW.MUNICHRE.COM, Swiss Re WWW.SWISSRE.COM and Lloyd's WWW.LLOYDS.COM give more recent information about the problems the reinsurance industry has to face.

The philosophy of classical non-life insurance is mainly based on the idea that large claims in a large portfolio have less influence and are "averaged out" by virtue of the strong law of large numbers and the central limit theorem. Over the last few years, extremely large claims have hit the reinsurance industry. Those include the claims which are summarized in Table 3.2.18. In order to deal with those claims, averaging techniques are insufficient; the expectation and the variance of a claim size sample tells one very little about the largest claims in the portfolio. Similar observations have been made in climatology, hydrology and meteorology: extreme events are not described by the normal distribution and its parameters. In those areas special techniques have been developed to deal with extremes. They run under the name of extreme value theory and extreme value statistics. We refer to the monograph Embrechts et al. [29] and the references therein for a comprehensive treatment of these topics.

Exercises

(1) An *extreme value distribution* F satisfies the following property: for every $n \geq 1$ there exist constants $c_n > 0$ and $d_n \in \mathbb{R}$ such that for iid random variables X_i with common distribution F,

$$c_n^{-1}(X_1 + \cdots + X_n - d_n) \overset{d}{=} X_1.$$

(a) Verify that the Gumbel distribution with distribution function $\Lambda(x) = e^{-e^{-x}}$, $x \in \mathbb{R}$, the Fréchet distribution with distribution function $\Phi_\alpha(x) = \exp\{-x^{-\alpha}\}$, $x > 0$, for some $\alpha > 0$, and the Weibull distribution with distribution function $\Psi_\alpha(x) = \exp\{-|x|^\alpha\}$, $x < 0$, for some $\alpha > 0$, are extreme value distributions. It can be shown that, up to changes of shift and location, these three distributions are the only extreme value distributions.

(b) The extreme value distributions are known to be the only non-degenerate limit distributions for partial maxima $M_n = \max(X_1, \ldots, X_n)$ of iid random variables X_i after suitable scaling and centering, i.e., there exist $c_n > 0$ and $d_n \in \mathbb{R}$ such that

$$c_n^{-1}(M_n - d_n) \overset{d}{\to} Y \sim H \in \{\Lambda, \Phi_\alpha, \Psi_\alpha\}. \tag{3.4.50}$$

Find suitable constants $c_n > 0$, $d_n \in \mathbb{R}$ and extreme value distributions H such that (3.4.50) holds for (i) Pareto, (ii) exponentially distributed, (iii) uniformly distributed claim sizes.

4

Ruin Theory

In Chapter 3 we studied the distribution and some distributional characteristics of the total claim amount $S(t)$ for fixed t as well as for $t \to \infty$. Although we sometimes used the structure of $S = (S(t))_{t \geq 0}$ as a stochastic process, for example of the renewal model, we did not really investigate the *finite-dimensional distributions* of the process S or any functional of S on a finite interval $[0, T]$ or on the interval $[0, \infty)$. Early on, with the path-breaking work of Crámer [23], the so-called *ruin probability* was introduced as a measure of risk which takes into account the temporal aspect of the insurance business over a finite or infinite time horizon. It is the aim of this section to report about Cramér's ruin bound and to look at some extensions. We start in Section 4.1 by introducing the basic notions related to ruin, including the net profit condition and the risk process. In Section 4.2 we collect some bounds on the probability of ruin. Those include the famous Lundberg inequality and Cramér's fundamental result in the case of small claim sizes. We also consider the large claim case. It turns out that the large and the small claim case lead to completely different bounds for ruin probabilities. In the small claim case ruin occurs as a collection of "atypical" claim sizes, whereas in the large claim case ruin happens as the result of one large claim size.

4.1 Risk Process, Ruin Probability and Net Profit Condition

Throughout this section we consider the total claim amount process

$$S(t) = \sum_{i=1}^{N(t)} X_i, \quad t \geq 0,$$

in the renewal model. This means that the iid sequence (X_i) of positive claim sizes with common distribution function F is independent of the claim arrival sequence (T_n) given by the renewal sequence

$$T_0 = 0, \quad T_n = W_1 + \cdots + W_n, \quad n \geq 1,$$

where the positive inter-arrival times W_n are assumed to be iid. Then the claim number process

$$N(t) = \#\{n \geq 1 : T_n \leq t\}, \quad t \geq 0,$$

is a renewal process which is independent of the claim size sequence (X_i).

In what follows we assume a continuous *premium income* $p(t)$ in the homogeneous portfolio which is described by the renewal model. We also assume for simplicity that p is a deterministic function and even *linear*:

$$p(t) = ct.$$

We call $c > 0$ the *premium rate*. The *surplus* or *risk process* of the portfolio is then defined by

$$U(t) = u + p(t) - S(t), \quad t \geq 0.$$

The quantity $U(t)$ is nothing but the insurer's capital balance at a given time t, and the process $U = (U(t))_{t \geq 0}$ describes the cashflow in the portfolio over time. The function $p(t)$ describes the inflow of capital into the business by time t and $S(t)$ describes the outflow of capital due to payments for claims occurred in $[0, t]$. If $U(t)$ is positive, the company has gained capital, if $U(t)$ is negative it has lost capital. The constant value $U(0) = u > 0$ is called *initial capital*. It is not further specified, but usually supposed to be a "huge" value.[1] Later on, the large size of u will be indicated by taking limits as $u \to \infty$.

In the top graph of Figure 4.1.2 we see an idealized path of the process U. The process U starts at the initial capital u. Then the path increases linearly with slope c until time $T_1 = W_1$, when the first claim happens. The process decreases by the size X_1 of the first claim. In the interval $[T_1, T_2)$ the process again increases with slope c until a second claim occurs at time T_2, when it jumps downward by the amount of X_2, etc. In the figure we have also indicated that negative values are possible for $U(t)$ if there is a sufficiently large claim X_i which pulls the path of U below zero. The event that U ever falls below zero is called *ruin*.

Definition 4.1.1 (Ruin, ruin time, ruin probability)
The event that U ever falls below zero is called ruin:

$$\text{Ruin} = \{U(t) < 0 \quad \text{for some } t > 0\}.$$

---

[1] The assumption of a large initial capital is not just a mathematical assumption but also an economic necessity, which is reinforced by the supervisory authorities. In any civilized country it is not possible to start up an insurance business without a sufficiently large initial capital (reserve), which prevents the business from bankruptcy due to too many small or a few large claim sizes in the first period of its existence, before the premium income can balance the losses and the gains.

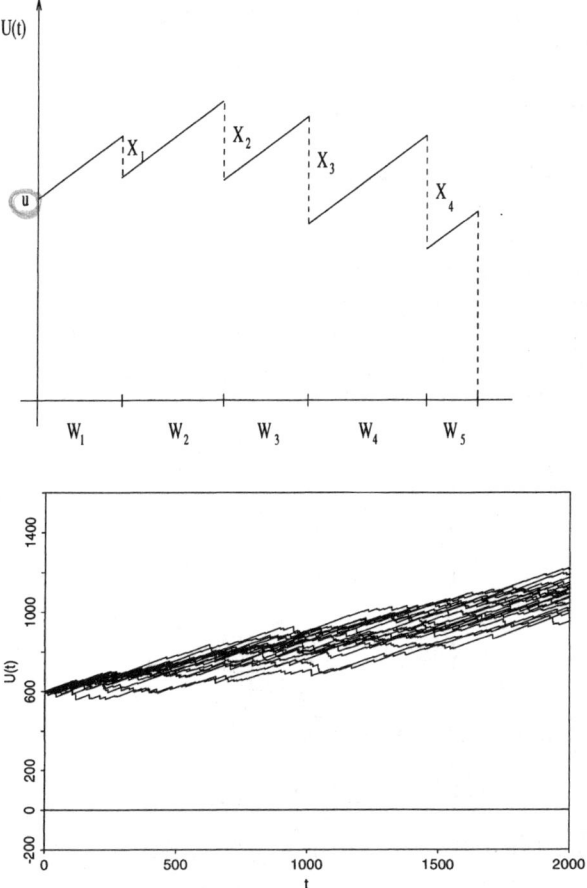

Figure 4.1.2 Top: *An idealized realization of the risk process* U. *Bottom: Some realizations of the risk process* U *for exponential claim sizes and a homogeneous Poisson claim number process* N. *Ruin does not occur in this graph: all paths stay positive.*

The time T when the process falls below zero for the first time is called ruin time:

$$T = \inf \{t > 0 : U(t) < 0\}\,.$$

The probability of ruin is then given by

$$\psi(u) = P(\text{Ruin} \mid U(0) = u) = P(T < \infty)\,, \quad u > 0\,. \qquad (4.1.1)$$

In the definition we made use of the fact that

$$\text{Ruin} = \bigcup_{t \geq 0} \{U(t) < 0\} = \left\{\inf_{t \geq 0} U(t) < 0\right\} = \{T < \infty\}\,.$$

The random variable T is not necessarily real-valued. Depending on the conditions on the renewal model, T may assume the value ∞ with positive probability. In other words, T is an *extended random variable*.

Both the event of ruin and the ruin time depend on the initial capital u, which we often suppress in the notation. The condition $U(0) = u$ in the ruin probability in (4.1.1) is artificial since $U(0)$ is a constant. This "conditional probability" is often used in the literature in order to indicate what the value of the initial capital is.

By construction of the risk process U, ruin can occur only at the times $t = T_n$ for some $n \geq 1$, since U linearly increases in the intervals $[T_n, T_{n+1})$. We call the sequence $(U(T_n))$ the *skeleton process* of the risk process U. Using the skeleton process, we can express ruin in terms of the inter-arrival times W_n, the claim sizes X_n and the premium rate c.

$$\text{Ruin} = \left\{ \inf_{t>0} U(t) < 0 \right\} = \left\{ \inf_{n \geq 1} U(T_n) < 0 \right\}$$

$$= \left\{ \inf_{n \geq 1} \left[u + p(T_n) - S(T_n) \right] < 0 \right\}$$

$$= \left\{ \inf_{n \geq 1} \left[u + cT_n - \sum_{i=1}^{n} X_i \right] < 0 \right\} .$$

In the latter step we used the fact that

$$N(T_n) = \#\{i \geq 1 : T_i \leq T_n\} = n \quad \text{a.s.}$$

since we assumed that $W_j > 0$ a.s. for all $j \geq 1$. Write

$$Z_n = X_n - cW_n , \quad S_n = Z_1 + \cdots + Z_n , \quad n \geq 1 , \quad S_0 = 0 .$$

Then we have the following alternative expression for the ruin probability $\psi(u)$ with initial capital u:

$$\psi(u) = P\left(\inf_{n \geq 1} (-S_n) < -u \right) = P\left(\sup_{n \geq 1} S_n > u \right) . \tag{4.1.2}$$

Since each of the sequences (W_i) and (X_i) consists of iid random variables and the two sequences are mutually independent, the ruin probability $\psi(u)$ is nothing but the tail probability of the supremum functional of the random walk (S_n). It is clear by its construction that this probability is not easily evaluated since one has to study a very complicated functional of a sophisticated random process. Nevertheless, the ruin probability has attracted enormous attention in the literature on applied probability theory. In particular, the asymptotic behavior of $\psi(u)$ as $u \to \infty$ has been of interest. The quantity $\psi(u)$ is a complex measure of the global behavior of an insurance portfolio as time goes by. The main aim is to avoid ruin with probability 1, and the probability that the random walk (S_n) exceeds the high threshold u should be so small that the event of ruin can be excluded from any practical considerations if the initial capital u is sufficiently large.

Since we are dealing with a random walk (S_n) we expect that we can conclude, from certain asymptotic results for the sample paths of (S_n), some elementary properties of the ruin probability. In what follows, we assume that both EW_1 and EX_1 are finite. This is a weak regularity condition on the inter-arrival times and the claim sizes which is met in most cases of practical interest. But then we also know that $EZ_1 = EX_1 - cEW_1$ is well-defined and finite. The random walk (S_n) satisfies the strong law of large numbers:

$$\frac{S_n}{n} \overset{\text{a.s.}}{\to} EZ_1 \quad \text{as } n \to \infty,$$

which in particular implies that $S_n \overset{\text{a.s.}}{\to} +\infty$ or $-\infty$ a.s. according to whether EZ_1 is positive or negative. Hence if $EZ_1 > 0$, ruin is unavoidable whatever the initial capital u.

If $EZ_1 = 0$ it follows from some deep theory on random walks (e.g. Spitzer [74]) that for a.e. ω there exists a subsequence $(n_k(\omega))$ such that $S_{n_k}(\omega) \to \infty$ and another subsequence $(m_k(\omega))$ such that $S_{m_k}(\omega) \overset{\text{a.s.}}{\to} -\infty$. Hence $\psi(u) = 1$ in this case as well.[2]

In any case, we may conclude the following:

Proposition 4.1.3 (Ruin with probability 1)
If EW_1 and EX_1 are finite and the condition

$$EZ_1 = EX_1 - cEW_1 \geq 0 \tag{4.1.3}$$

holds then, for every fixed $u > 0$, ruin occurs with probability 1.

From Proposition 4.1.3 we learn that any insurance company should choose the premium $p(t) = ct$ in such a way that $EZ_1 < 0$. This is the only way to avoid ruin occurring with probability 1. If $EZ_1 < 0$ we may hope that $\psi(u)$ is different from 1.

Because of its importance we give a special name to the converse of condition (4.1.3).

Definition 4.1.4 (Net profit condition)
We say that the renewal model satisfies the net profit condition (NPC) *if*

$$EZ_1 = EX_1 - cEW_1 < 0. \tag{4.1.4}$$

---

[2] Under the stronger assumptions $EZ_1 = 0$ and $\text{var}(Z_1) < \infty$ one can show that the multivariate central limit theorem implies $\psi(u) = 1$ for every $u > 0$; see Exercise 1 on p. 160.

The interpretation of the NPC is rather intuitive. In a given unit of time the expected claim size EX_1 has to be smaller than the premium income in this unit of time, represented by the expected premium $c\,EW_1$. In other words, the average cashflow in the portfolio is on the positive side: on average, more premium flows into the portfolio than claim sizes flow out. This does not mean that ruin is avoided since the expectation of a stochastic process says relatively little about the fluctuations of the process.

Example 4.1.5 (NPC and premium calculation principle)
The relation of the NPC with the premium calculation principles mentioned in Section 3.1.3 is straightforward. For simplicity, assume the Cramér-Lundberg model; see p. 18. We know that

$$ES(t) = EN(t)\,EX_1 = \lambda t\,EX_1 = \frac{EX_1}{EW_1}\,t\,.$$

If we choose the premium $p(t) = ct$ with $c = EX_1/EW_1$, we are in the net premium calculation principle. In this case, $EZ_1 = 0$, i.e., ruin is unavoidable with probability 1. This observation supports the intuitive argument against the net principle we gave in Section 3.1.3.

Now assume that we have the expected value or the variance premium principle. Then for some positive safety loading ρ,

$$p(t) = (1+\rho)\,ES(t) = (1+\rho)\,\frac{EX_1}{EW_1}\,t\,.$$

This implies the premium rate

$$c = (1+\rho)\,\frac{EX_1}{EW_1}\,. \tag{4.1.5}$$

In particular, $EZ_1 < 0$, i.e., the NPC is satisfied. $\square$

Exercises

(1) We know that the ruin probability $\psi(u)$ in the renewal model has representation

$$\psi(u) = P\left(\sup_{n \geq 1} S_n > u\right), \tag{4.1.6}$$

where $S_n = Z_1 + \cdots + Z_n$ is a random walk with iid step sizes $Z_i = X_i - c\,W_i$. Assume that the conditions $EZ_1 = 0$ and $\mathrm{var}(Z_1) < \infty$ hold.
(a) Apply the central limit theorem to show that

$$\lim_{u \to \infty} \psi(u) \geq 1 - \Phi(0) = 0.5\,,$$

where Φ is the standard normal distribution function. Hint: Notice that $\psi(u) \geq P(S_n > u)$ for every $n \geq 1$.

(b) Let (Y_n) be an iid sequence of standard normal random variables. Show that for every $n \geq 1$,

$$\lim_{u \to \infty} \psi(u) \geq P\left(\max\left(Y_1, Y_1 + Y_2, \ldots, Y_1 + \cdots + Y_n\right) \geq 0\right).$$

Hint: Apply the multivariate central limit theorem and the continuous mapping theorem; see for example Billingsley [12].

(c) Standard *Brownian motion* $(B_t)_{t \geq 0}$ is a stochastic process with independent stationary increments and continuous sample paths, starts at zero, i.e., $B_0 = 0$ a.s., and $B_t \sim N(0, t)$ for $t \geq 0$. Show that

$$\lim_{u \to \infty} \psi(u) \geq P\left(\max_{0 \leq s \leq 1} B_s \geq 0\right).$$

Hint: Use (b).

(d) It is a well-known fact (see, for example, Resnick [65], Corollary 6.5.3 on p. 499) that Brownian motion introduced in (c) satisfies the reflection principle

$$P\left(\max_{0 \leq s \leq 1} B_s \geq x\right) = 2\, P(B_1 > x), \qquad x \geq 0.$$

Use this result and (c) to show that $\lim_{u \to \infty} \psi(u) = 1$.

(e) Conclude from (d) that $\psi(u) = 1$ for every $u > 0$. Hint: Notice that $\psi(u) \geq \psi(u')$ for $u \leq u'$.

(2) Consider the total claim amount process

$$S(t) = \sum_{i=1}^{N(t)} X_i, \qquad t \geq 0,$$

where (X_i) are iid positive claim sizes, independent of the Poisson process N with an a.e. positive and continuous intensity function λ. Choose the premium such that

$$p(t) = c \int_0^t \lambda(s)\, ds = c\,\mu(t),$$

for some premium rate $c > 0$ and consider the ruin probability

$$\psi(u) = P\left(\inf_{t \geq 0}(u + p(t) - S(t)) < 0\right),$$

for some positive initial capital u. Show that $\psi(u)$ coincides with the ruin probability in the Cramér-Lundberg model with Poisson intensity 1, initial capital u and premium rate c. Which condition is needed in order to avoid ruin with probability 1?

4.2 Bounds for the Ruin Probability

4.2.1 Lundberg's Inequality

In this section we derive an elementary upper bound for the ruin probability $\psi(u)$. We always assume the renewal model with the NPC (4.1.4). In addition, we assume a *small claim condition*: the existence of the moment generating function of the claim size distribution in a neighborhood of the origin

$$m_{X_1}(h) = E e^{h X_1}, \quad h \in (-h_0, h_0) \quad \text{for some } h_0 > 0. \qquad (4.2.7)$$

By Markov's inequality, for $h \in (0, h_0)$,

$$P(X_1 > x) \le e^{-h x} m_{X_1}(h) \quad \text{for all } x > 0.$$

Therefore $P(X_1 > x)$ decays to zero exponentially fast. We have learned in Section 3.2 that this condition is perhaps not the most realistic condition for real-life claim sizes, which often tend to have heavier tails, in particular, their moment generating function is not finite in any neighborhood of the origin. However, we present this material here for small claims since the classical work by Lundberg and Cramér was done under this condition.

The following notion will be crucial.

Definition 4.2.1 (Adjustment or Lundberg coefficient)
Assume that the moment generating function of Z_1 exists in some neighborhood $(-h_0, h_0)$, $h_0 > 0$, of the origin. If a unique positive solution r to the equation

$$m_{Z_1}(h) = E e^{h (X_1 - c W_1)} = 1 \qquad (4.2.8)$$

exists it is called the adjustment *or Lundberg coefficient.*

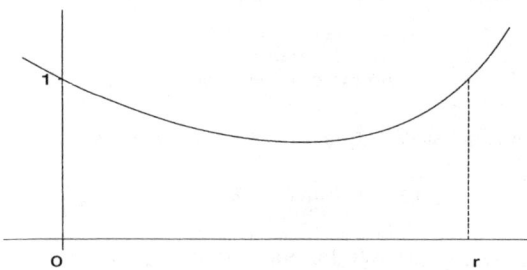

Figure 4.2.2 *A typical example of the function $f(h) = m_{Z_1}(h)$ with the Lundberg coefficient r.*

The existence of the moment generating function $m_{X_1}(h)$ for $h \in [0, h_0)$ implies the existence of $m_{Z_1}(h) = m_{X_1}(h) m_{c W_1}(-h)$ for $h \in [0, h_0)$ since $m_{c W_1}(-h) \le 1$ for all $h \ge 0$. For $h \in (-h_0, 0)$ the same argument implies that $m_{Z_1}(h)$ exists if $m_{c W_1}(-h)$ is finite. Hence the moment generating function of Z_1 exists in a neighborhood of zero if the moment generating functions of X_1

and cW_1 do. In the Cramér-Lundberg model with intensity λ for the claim number process N, $m_{cW_1}(h) = \lambda/(\lambda - ch)$ exists for $h < \lambda/c$.

In Definition 4.2.1 it was implicitly mentioned that r is unique, provided it exists as the solution to (4.2.8). The uniqueness can be seen as follows. The function $f(h) = m_{Z_1}(h)$ has derivatives of all orders in $(-h_0, h_0)$. This is a well-known property of moment generating functions. Moreover, $f'(0) = EZ_1 < 0$ by the NPC and $f''(h) = E(Z_1^2 \exp\{hZ_1\}) > 0$ since $Z_1 \neq 0$ a.s. The condition $f'(0) < 0$ and continuity of f imply that f decreases in some neighborhood of zero. On the other hand, $f''(h) > 0$ implies that f is convex. This implies that, if there exists some $h_c \in (0, h_0)$ such that $f'(h_c) = 0$, then f changes its monotonicity behavior from decrease to increase at h_c. For $h > h_c$, f increases; see Figure 4.2.2 for some illustration. Therefore the solution r of the equation $f(h) = 1$ is unique, provided the moment generating function exists in a sufficiently large neighborhood of the origin. A sufficient condition for this to happen is that there exists $0 < h_1 \leq \infty$ such that $f(h) < \infty$ for $h < h_1$ and $\lim_{h \uparrow h_1} f(h) = \infty$. This means that the moment generating function $f(h)$ increases continuously to infinity. In particular, it assumes the value 1 for sufficiently large h.

From this argument we also see that the existence of the adjustment coefficient as the solution to (4.2.8) is not automatic; the existence of the moment generating function of Z_1 in some neighborhood of the origin is not sufficient to ensure that there is some $r > 0$ with $f(r) = 1$.

Now we are ready to formulate one of the classical results in insurance mathematics.

Theorem 4.2.3 (The Lundberg inequality)
Assume the renewal model with NPC (4.1.4). Also assume that the adjustment coefficient r exists. Then the following inequality holds for all $u > 0$:

$$\psi(u) \leq e^{-ru}.$$

The exponential bound of the Lundberg inequality ensures that the probability of ruin is very small if one starts with a large initial capital u. Clearly, the bound also depends on the magnitude of the adjustment coefficient. The smaller r is, the more risky is the portfolio. In any case, the result tells us that, under a small claim condition and with a large initial capital, there is *in principle* no danger of ruin in the portfolio. We will see later in Section 4.2.4 that this statement is incorrect for portfolios with large claim sizes. We also mention that this result is much more informative than we ever could derive from the average behavior of the portfolio given by the strong law of large numbers for $S(t)$ supplemented by the central limit theorem for $S(t)$.

Proof. We will prove the Lundberg inequality by induction. Write

$$\psi_n(u) = P\left(\max_{1 \leq k \leq n} S_k > u\right) = P(S_k > u \text{ for some } k \in \{1, \ldots, n\})$$

and notice that $\psi_n(u) \uparrow \psi(u)$ as $n \to \infty$ for every $u > 0$. Thus it suffices to prove that

$$\psi_n(u) \le e^{-ru} \quad \text{for all } n \ge 1 \text{ and } u > 0. \tag{4.2.9}$$

We start with $n = 1$. By Markov's inequality and the definition of the adjustment coefficient,

$$\psi_1(u) \le e^{-ru} \, m_{Z_1}(r) = e^{-ru}.$$

This proves (4.2.9) for $n = 1$. Now assume that (4.2.9) holds for $n = k \ge 1$. In the induction step we use a typical renewal argument. Write F_{Z_1} for the distribution function of Z_1. Then

$$\psi_{k+1}(u) = P\left(\max_{1 \le n \le k+1} S_n > u\right)$$

$$= P(Z_1 > u) + P\left(\max_{2 \le n \le k+1} (Z_1 + (S_n - Z_1)) > u, Z_1 \le u\right)$$

$$= \int_{(u,\infty)} dF_{Z_1}(x) + \int_{(-\infty,u]} P\left(\max_{1 \le n \le k}[x + S_n] > u\right) dF_{Z_1}(x)$$

$$= p_1 + p_2.$$

We consider p_2 first. Using the induction assumption for $n = k$, we have

$$p_2 = \int_{(-\infty,u]} P\left(\max_{1 \le n \le k} S_n > u - x\right) dF_{Z_1}(x) = \int_{(-\infty,u]} \psi_k(u - x) \, dF_{Z_1}(x)$$

$$\le \int_{(-\infty,u]} e^{r(x-u)} \, dF_{Z_1}(x).$$

Similarly, by Markov's inequality,

$$p_1 \le \int_{(u,\infty)} e^{r(x-u)} \, dF_{Z_1}(x).$$

Hence, by the definition of the adjustment coefficient r,

$$p_1 + p_2 \le e^{-ru} \, m_{Z_1}(r) = e^{-ru},$$

which proves (4.2.9) for $n = k + 1$ and concludes the proof. $\square$

Next we give a benchmark example for the Lundberg inequality.

Example 4.2.4 (Lundberg inequality for exponential claims)
Consider the Cramér-Lundberg model with iid exponential $\text{Exp}(\gamma)$ claim sizes and Poisson intensity λ. This means in particular that the W_i's are iid exponential $\text{Exp}(\lambda)$ random variables. The moment generating function of an $\text{Exp}(a)$ distributed random variable A is given by

$$m_A(h) = \frac{a}{a-h}, \quad h < a.$$

Hence the moment generating function of $Z_1 = X_1 - cW_1$ takes the form

$$m_{Z_1}(h) = m_{X_1}(h)\, m_{cW_1}(-h) = \frac{\gamma}{\gamma - h}\frac{\lambda}{\lambda + ch}, \quad -\lambda/c < h < \gamma.$$

The adjustment coefficient is then the solution to the equation

$$1 + h\frac{c}{\lambda} = \frac{\gamma}{\gamma - h} = \frac{1}{1 - h\,EX_1}, \tag{4.2.10}$$

where $\gamma = (EX_1)^{-1}$. Now recall that the NPC holds:

$$\frac{EX_1}{EW_1} = \frac{\lambda}{\gamma} < c$$

Under this condition, straightforward calculation shows that equation (4.2.10) has a unique positive solution given by

$$r = \gamma - \frac{\lambda}{c} > 0.$$

In Example 4.1.5 we saw that we can interpret the premium rate c in terms of the expected value premium calculation principle:

$$c = \frac{EX_1}{EW_1}(1+\rho) = \frac{\lambda}{\gamma}(1+\rho).$$

Thus, in terms of the safety loading ρ,

$$r = \gamma\frac{\rho}{1+\rho}. \tag{4.2.11}$$

We summarize: In the Cramér-Lundberg model with iid $\mathrm{Exp}(\gamma)$ distributed claim sizes and Poisson intensity λ, the Lundberg inequality for the ruin probability $\psi(u)$ is of the form

$$\psi(u) \le \exp\left\{-\gamma\frac{\rho}{1+\rho}u\right\}, \quad u > 0. \tag{4.2.12}$$

From this inequality we get the intuitive meaning of the ruin probability $\psi(u)$ as a risk measure: ruin is very unlikely if u is large. However, the Lundberg bound is the smaller the larger we choose the safety loading ρ since $\rho/(1+\rho) \uparrow 1$ as $\rho \uparrow \infty$. The latter limit relation also tells us that the bound does not change significantly if ρ is sufficiently large. The right-hand side of (4.2.12) is also influenced by $\gamma = (EX_1)^{-1}$: the smaller the expected claim size, the smaller the ruin probability.

We will see in Example 4.2.13 that (4.2.12) is an almost precise estimate for $\psi(u)$ in the case of exponential claims: $\psi(u) = C\,\exp\{-u\,\gamma\,\rho/(1+\rho)\}$ for some positive C. $\qquad\square$

Comments

It is in general difficult, if not impossible, to determine the adjustment coefficient r as a function of the distributions of the claim sizes and the inter-arrival times. A few well-known examples where one can determine r explicitly can be found in Asmussen [4] and Rolski et al. [67]. In general, one depends on numerical or Monte Carlo approximations to r.

4.2.2 Exact Asymptotics for the Ruin Probability: the Small Claim Case

In this section we consider the Cramér-Lundberg model, i.e., the renewal model with a homogeneous Poisson process with intensity λ as claim number process. It is our aim to get bounds on the ruin probability $\psi(u)$ from above and from below.

The following result is one of the most important results of risk theory, due to Cramér [23].

Theorem 4.2.5 (Cramér's ruin bound)
Consider the Cramér-Lundberg model with NPC (4.1.4). In addition, assume that the claim size distribution function F_{X_1} has a density, the moment generating function of X_1 exists in some neighborhood $(-h_0, h_0)$ of the origin, the adjustment coefficient (see (4.2.8)) exists and lies in $(-h_0, h_0)$. Then there exists a constant $C > 0$ such that

$$\lim_{u \to \infty} e^{r u}\, \psi(u) = C\,.$$

The value of the constant C is given in (4.2.25). It involves the adjustment coefficient r, the expected claim size EX_1 and other characteristics of F_{X_1} as well as the safety loading ρ. We have chosen to express the NPC by means of ρ; see (4.1.5):

$$\rho = c\,\frac{EW_1}{EX_1} - 1 > 0\,.$$

The proof of this result is rather technical. In what follows, we indicate some of the crucial steps in the proof. We introduce some additional notation. The non-ruin probability is given by

$$\varphi(u) = 1 - \psi(u)\,.$$

As before, we write F_A for the distribution function of any random variable A and $\overline{F}_A = 1 - F_A$ for its tail.

The following auxiliary result is key to Theorem 4.2.5.

Lemma 4.2.6 (Fundamental integral equation for the non-ruin probability)
Consider the Cramér-Lundberg model with NPC and $EX_1 < \infty$. In addition,

assume that the claim size distribution function F_{X_1} has a density. Then the non-ruin probability $\varphi(u)$ satisfies the integral equation

$$\varphi(u) = \varphi(0) + \frac{1}{(1+\rho)\,EX_1} \int_0^u \overline{F}_{X_1}(y)\,\varphi(u-y)\,dy. \qquad (4.2.13)$$

Remark 4.2.7 Write

$$F_{X_1,I}(y) = \frac{1}{EX_1} \int_0^y \overline{F}_{X_1}(z)\,dz, \quad y > 0.$$

for the *integrated tail distribution function* of X_1. Notice that $F_{X_1,I}$ is indeed a distribution function since for any positive random variable A we have $EA = \int_0^\infty \overline{F}_A(y)\,dy$ and, therefore, $F_{X_1,I}(y) \uparrow 1$ as $y \uparrow \infty$. Now one can convince oneself that (4.2.13) takes the form

$$\varphi(u) = \varphi(0) + \frac{1}{1+\rho} \int_0^u \varphi(u-y)\,dF_{X_1,I}(y), \qquad (4.2.14)$$

which reminds one of a renewal equation; see (2.2.40). Recall that in Section 2.2.2 we considered some renewal theory. It will be the key to the bound of Theorem 4.2.5. □

Remark 4.2.8 The constant $\varphi(0)$ in (4.2.13) can be evaluated. Observe that $\varphi(u) \uparrow 1$ as $u \to \infty$. This is a consequence of the NPC and the fact that $S_n \to -\infty$ a.s., hence $\sup_{n\geq 1} S_n < \infty$ a.s. By virtue of (4.2.14) and the monotone convergence theorem,

$$1 = \lim_{u\uparrow\infty} \varphi(u) = \varphi(0) + \frac{1}{1+\rho} \lim_{u\uparrow\infty} \int_0^\infty I_{\{y\leq u\}}\,\varphi(u-y)\,dF_{X_1,I}(y)$$

$$= \varphi(0) + \frac{1}{1+\rho} \int_0^\infty 1\,dF_{X_1,I}(y)$$

$$= \varphi(0) + \frac{1}{1+\rho}.$$

Hence $\varphi(0) = \rho\,(1+\rho)^{-1}$. □

We continue with the proof of Lemma 4.2.6.

Proof. We again use a renewal argument. Recall from (4.1.2) that

$$\psi(u) = P\left(\sup_{n\geq 1} S_n > u\right) = 1 - \varphi(u),$$

where (S_n) is the random walk generated from the iid sequence (Z_n) with $Z_n = X_n - c\,W_n$. Then

$$\varphi(u) = P\left(\sup_{n\geq 1} S_n \leq u\right) = P\left(S_n \leq u \text{ for all } n \geq 1\right) \qquad (4.2.15)$$

$$= P\left(Z_1 \le u, S_n - Z_1 \le u - Z_1 \text{ for all } n \ge 2\right)$$

$$= E\left[I_{\{Z_1 \le u\}}\, P\left(S_n - Z_1 \le u - Z_1 \text{ for all } n \ge 2 \mid Z_1\right)\right]$$

$$= \int_{w=0}^{\infty}\int_{x=0}^{u+cw} P\left(S_n - Z_1 \le u - (x - cw) \text{ for all } n \ge 2\right) dF_{X_1}(x)\, dF_{W_1}(w)$$

$$= \int_{w=0}^{\infty}\int_{x=0}^{u+cw} P\left(S_n \le u - (x - cw) \text{ for all } n \ge 1\right) dF_{X_1}(x)\, \lambda e^{-\lambda w}\, dw.$$

$$(4.2.16)$$

Here we used the independence of $Z_1 = X_1 - cW_1$ and the sequence $(S_n - Z_1)_{n\ge 2}$. This sequence has the same distribution as $(S_n)_{n\ge 1}$, and the random variable W_1 has $\text{Exp}(\lambda)$ distribution. An appeal to (4.2.15) and (4.2.16) yields

$$\varphi(u) = \int_{w=0}^{\infty}\int_{x=0}^{u+cw} \varphi(u - x + cw)\, dF_{X_1}(x)\, \lambda e^{-\lambda w}\, dw.$$

With the substitution $z = u + cw$ we arrive at

$$\varphi(u) = \frac{\lambda}{c} e^{u\lambda/c} \int_{z=u}^{\infty} e^{-\lambda z/c} \int_{x=0}^{z} \varphi(z - x)\, dF_{X_1}(x)\, dz. \qquad (4.2.17)$$

Since we assumed that F_{X_1} has a density, the function

$$g(z) = \int_0^z \varphi(z - x)\, dF_{X_1}(x)$$

is continuous. By virtue of (4.2.17),

$$\varphi(u) = \frac{\lambda}{c} e^{u\lambda/c} \int_{z=u}^{\infty} e^{-\lambda z/c}\, g(z)\, dz,$$

and, hence, φ is even differentiable. Differentiating (4.2.17), we obtain

$$\varphi'(u) = \frac{\lambda}{c}\, \varphi(u) - \frac{\lambda}{c} \int_0^u \varphi(u - x)\, dF_{X_1}(x).$$

Now integrate the latter identity and apply partial integration:

$$\varphi(t) - \varphi(0) - \frac{\lambda}{c} \int_0^t \varphi(u)\, du$$

$$= -\frac{\lambda}{c} \int_0^t \int_0^u \varphi(u - x)\, dF_{X_1}(x)\, du$$

$$= -\frac{\lambda}{c} \int_0^t \left[\varphi(u - x)\, F_{X_1}(x)\Big|_0^u + \int_0^u \varphi'(u - x)\, F_{X_1}(x)\, dx\right] du$$

$$= -\frac{\lambda}{c} \int_0^t \left[\varphi(0)\, F_{X_1}(u) + \int_0^u \varphi'(u - x)\, F_{X_1}(x)\, dx\right] du.$$

In the last step we used $F_{X_1}(0) = 0$ since $X_1 > 0$ a.s. Now interchange the integrals:

$$\varphi(t) - \varphi(0)$$

$$= \frac{\lambda}{c} \int_0^t \varphi(u)\, du - \frac{\lambda}{c} \varphi(0) \int_0^t F_{X_1}(u)\, du - \frac{\lambda}{c} \int_0^t F_{X_1}(x)\left[\varphi(t-x) - \varphi(0)\right] dx$$

$$= \frac{\lambda}{c} \int_0^t \varphi(t-u)\, du - \frac{\lambda}{c} \int_0^t F_{X_1}(x)\, \varphi(t-x)\, dx$$

$$= \frac{\lambda}{c} \int_0^t \overline{F}_{X_1}(x)\, \varphi(t-x)\, dx\,. \tag{4.2.18}$$

Observe that

$$\frac{\lambda}{c} = \frac{1}{1+\rho}\frac{1}{EX_1}\,,$$

see (4.1.5). The latter relation and (4.2.18) prove the lemma. $\qquad\square$

Lemma 4.2.6 together with Remarks 4.2.7 and 4.2.8 ensures that the non-ruin probability φ satisfies the equation

$$\varphi(u) = \frac{\rho}{1+\rho} + \frac{1}{1+\rho} \int_0^u \varphi(u-y)\, dF_{X_1,I}(y)\,, \tag{4.2.19}$$

where

$$F_{X_1,I}(x) = \frac{1}{EX_1} \int_0^x \overline{F}_{X_1}(y)\, dy\,, \quad x > 0\,,$$

is the integrated tail distribution function of the claim sizes X_i. Writing

$$q = \frac{1}{1+\rho}$$

and switching in (4.2.19) from $\varphi = 1 - \psi$ to ψ, we obtain the equation

$$\psi(u) = q\,\overline{F}_{X_1,I}(u) + \int_0^u \psi(u-x)\, d\left(q\, F_{X_1,I}(x)\right)\,. \tag{4.2.20}$$

This looks like a renewal equation, see (2.2.40):

$$R(t) = u(t) + \int_{[0,t]} R(t-y)\, dF(y)\,, \tag{4.2.21}$$

where F is the distribution function of a positive random variable, u is a function on $[0,\infty)$ bounded on every finite interval and R is an unknown function. However, there is one crucial difference between (4.2.20) and (4.2.21): in the former equation one integrates with respect to the measure $q\,F_{X_1,I}$ which is *not* a probability measure since $\lim_{x\to\infty}(q\,F_{X_1,I}(x)) = q < 1$. Therefore (4.2.20) is called a *defective renewal equation*. Before one can apply standard renewal theory, one has to transform (4.2.20) into the standard form (4.2.21) for some distribution function F.

Only at this point the notion of adjustment coefficient r comes into consideration. We define the distribution function $F^{(r)}$ for $x > 0$:

$$F^{(r)}(x) = \int_0^x e^{ry}\,d\,(q\,F_{X_1,I}(y)) = q\int_0^x e^{ry}\,dF_{X_1,I}(y)$$

$$= \frac{q}{EX_1}\int_0^x e^{ry}\,\overline{F}_{X_1}(y)\,dy\,.$$

The distribution generated by $F^{(r)}$ is said to be the *Esscher transform* or the *exponentially tilted distribution* of F. This is indeed a distribution function since $F^{(r)}(x)$ is non-decreasing and has a limit as $x \to \infty$ given by

$$\frac{q}{EX_1}\int_0^\infty e^{ry}\,\overline{F}_{X_1}(y)\,dy = 1\,. \qquad (4.2.22)$$

This identity can be shown by partial integration and the definition of the adjustment coefficient r. Verify (4.2.22); see also Exercise 3 on p. 182.

Multiplying both sides of (4.2.20) by e^{ru}, we obtain the equation

$$e^{ru}\,\psi(u) = q\,e^{ru}\,\overline{F}_{X_1,I}(u) + \int_0^u e^{r\,(u-x)}\,\psi(u-x)\,e^{rx}\,d\,(q\,F_{X_1,I}(x))$$

$$= q\,e^{ru}\,\overline{F}_{X_1,I}(u) + \int_0^u e^{r\,(u-x)}\,\psi(u-x)\,dF^{(r)}(x)\,, \qquad (4.2.23)$$

which is of renewal type (4.2.21) with $F = F^{(r)}$, $u(t) = q\,e^{rt}\,\overline{F}_{X_1,I}(t)$ and unknown function $R(t) = e^{rt}\,\psi(t)$. The latter function is bounded on finite intervals. Therefore we may apply Smith's key renewal Theorem 2.2.12(1) to conclude that the renewal equation (4.2.23) has solution

$$R(t) = e^{rt}\,\psi(t) = \int_{[0,t]} u(t-y)\,dm^{(r)}(y)$$

$$= q\int_{[0,t]} e^{r\,(t-y)}\,\overline{F}_{X_1,I}(t-y)\,dm^{(r)}(y)\,, \qquad (4.2.24)$$

where $m^{(r)}$ is the renewal function corresponding to the renewal process whose inter-arrival times have common distribution function $F^{(r)}$. In general, we do not know the function $m^{(r)}$. However, Theorem 2.2.12(2) gives us the asymptotic order of the solution to (4.2.23) as $u \to \infty$:

$$C = \lim_{u\to\infty} e^{ru}\,\psi(u) = \lambda\,q\int_0^\infty e^{ry}\,\overline{F}_{X_1,I}(y)\,dy\,.$$

For the application of Theorem 2.2.12(2) we would have to verify whether $u(t) = q\,e^{rt}\overline{F}_{X_1,I}(t)$ is directly Riemann integrable. We refer to p. 31 in Embrechts et al. [29] for an argument. Calculation yields

$$C = \left[\frac{r}{\rho\,EX_1}\int_0^\infty x\,e^{rx}\overline{F}_{X_1}(x)\,dx\right]^{-1}. \tag{4.2.25}$$

This finishes the proof of the Cramér ruin bound of Theorem 4.2.5 .   □

We mention in passing that the definition of the constant C in (4.2.25) requires more than the existence of the moment generating function $m_{X_1}(h)$ at $h = r$. This condition is satisfied since we assume that $m_{X_1}(h)$ exists in an open neighborhood of the origin, containing r.

Example 4.2.9 (The ruin probability in the Cramér-Lundberg model with exponential claim sizes)
As mentioned above, the solution (4.2.24) to the renewal equation for $e^{ru}\,\psi(u)$ is in general not explicitly given. However, if we assume that the iid claim sizes X_i are $\mathrm{Exp}(\gamma)$ for some $\gamma > 0$, then this solution can be calculated. Indeed, the exponentially tilted distribution function $F^{(r)}$ is then $\mathrm{Exp}(\gamma - r)$ distributed, where $\gamma - r = \gamma/(1 + \rho) = \gamma q$; see (4.2.11). Recall that the renewal function $m^{(r)}$ is given by $m^{(r)}(t) = EN^{(r)}(t) + 1$, where $N^{(r)}$ is the renewal process generated by the iid inter-arrival times $W_i^{(r)}$ with common distribution function $F^{(r)}$. Since $F^{(r)}$ is $\mathrm{Exp}(\gamma q)$, the renewal process $N^{(r)}$ is homogeneous Poisson with intensity γq and therefore

$$m^{(r)}(t) = \gamma\,q\,t + 1\,, \quad t > 0\,.$$

According to Theorem 2.2.12(1), we have to interpret the integral in (4.2.24) such that $m^{(r)}(y) = 0$ for $y < 0$. Taking the jump of $m^{(r)}$ at zero into account, (4.2.24) reads as follows:

$$e^{rt}\,\psi(t) = q\,e^{rt}e^{-\gamma t} + \gamma\,q^2\int_0^t e^{r(t-y)}\,e^{-\gamma(t-y)}\,dy$$

$$= q\,e^{-t(\gamma-r)} + \gamma\,q^2\,\frac{1}{\gamma-r}\left(1 - e^{-t(\gamma-r)}\right)$$

$$= q\,.$$

This means that one gets the exact ruin probability $\psi(t) = q\,e^{-rt}$.   □

Example 4.2.10 (The tail of the distribution of the solution to a stochastic recurrence equation)
The following model has proved useful in various applied contexts:

$$Y_t = A_t\,Y_{t-1} + B_t\,, \quad t \in \mathbb{Z}\,, \tag{4.2.26}$$

where A_t and B_t are random variables, possibly dependent for each t, and the sequence of pairs (A_t, B_t) constitutes an iid sequence. Various popular models for financial *log-returns*[3] are closely related to the *stochastic recurrence equation* (4.2.26). For example, consider an *autoregressive conditionally heteroscedastic process of order 1* (ARCH(1))

$$X_t = \sigma_t Z_t, \quad t \in \mathbb{Z},$$

where (Z_t) is an iid sequence with unit variance and mean zero.[4] The *squared volatility sequence* (σ_t^2) is given by the relation

$$\sigma_t^2 = \alpha_0 + \alpha_1 X_{t-1}^2, \quad t \in \mathbb{Z},$$

where α_0, α_1 are positive constants. Notice that $Y_t = X_t^2$ satisfies the stochastic recurrence equation (4.2.26) with $A_t = \alpha_1 Z_t^2$ and $B_t = \alpha_0 Z_t^2$:

$$X_t^2 = (\alpha_0 + \alpha_1 X_{t-1}^2) Z_t^2 = [\alpha_1 Z_t^2] X_{t-1}^2 + [\alpha_0 Z_t^2]. \quad (4.2.27)$$

An extension of the ARCH(1) model is the GARCH(1,1) model (*generalized ARCH model of order* $(1,1)$) given by the equation

$$X_t = \sigma_t Z_t, \quad \sigma_t^2 = \alpha_0 + \alpha_1 X_{t-1}^2 + \beta_1 \sigma_{t-1}^2, \quad t \in \mathbb{Z}.$$

Here (Z_t) is again an iid sequence with mean zero and unit variance, and α_0, α_1 and β_1 are positive constants. The squared log-return series (X_t^2) does not satisfy a stochastic recurrence equation of type (4.2.26). However, the squared volatility sequence (σ_t^2) satisfies such an equation with $A_t = \alpha_1 Z_{t-1}^2 + \beta_1$ and $B_t = \alpha_0$:

$$\sigma_t^2 = \alpha_0 + \alpha_1 \sigma_{t-1}^2 Z_{t-1}^2 + \beta_1 \sigma_{t-1}^2 = \alpha_0 + [\alpha_1 Z_{t-1}^2 + \beta_1] \sigma_{t-1}^2.$$

In an insurance context, equation (4.2.26) has interpretation as present value of future accumulated payments which are subject to stochastic discounting. At the instants of time $t = 0, 1, 2, \ldots$ a payment B_t is made. Previous payments Y_{t-1} are discounted by the stochastic discount factor A_t, i.e., A_t^{-1} is the interest paid for one price unit in the tth period, for example, in year t. Then $Y_t = A_t Y_{t-1} + B_t$ is the present value of the payments after t time steps.

In what follows, we assume that (A_t) is an iid sequence of positive random variables and, for the ease of presentation, we only consider the case $B_t \equiv 1$. It is convenient to consider all sequences with index set $\mathbb{Z}$. Iteration of equation (4.2.26) yields

---

[3] For a price P_t of a risky asset (share price of stock, composite stock index, foreign exchange rate,...) which is reported at the times $t = 0, 1, 2, \ldots$ the log-differences $R_t = \log P_t - \log P_{t-1}$ constitute the log-returns. In contrast to the prices P_t, it is believed that the sequence (R_t) can be modeled by a stationary process.

[4] The sequence (Z_t) is often supposed to be iid standard normal.

$$Y_t = A_t A_{t-1} Y_{t-2} + A_t + 1$$
$$= A_t A_{t-1} A_{t-2} Y_{t-2} + A_t A_{t-1} + A_t + 1$$
$$\vdots \qquad\qquad \vdots \qquad\qquad \vdots$$
$$= A_t \cdots A_1 Y_0 + \sum_{i=1}^{t-1} \prod_{j=i+1}^{t} A_j + 1 .$$

The natural question arises as to whether "infinite iteration" yields anything useful, i.e., as to whether the sequence (Y_t) has series representation

$$Y_t = 1 + \sum_{i=-\infty}^{t-1} \prod_{j=i+1}^{t} A_j , \quad t \in \mathbb{Z}. \qquad (4.2.28)$$

Since we deal with an infinite series we first have to study its convergence behavior; this means we have to consider the question of its existence. If $E \log A_1$ is well-defined, the strong law of large numbers yields

$$|t - i|^{-1} T_{i,t} = |t - i|^{-1} \sum_{j=i+1}^{t} \log A_j \overset{\text{a.s.}}{\to} E \log A_1 \quad \text{as } i \to -\infty.$$

Now assume that $-\infty \le E \log A_1 < 0$ and choose $c \in (0,1)$ such that $E \log A_1 < \log c < 0$. Then the strong law of large numbers implies

$$\prod_{j=i+1}^{t} A_j = \exp\left\{|t-i|\left[|t-i|^{-1} T_{i,t}\right]\right\} \le \exp\left\{|t-i| \log c\right\} = c^{|t-i|}$$

for $i \le i_0 = i_0(\omega)$, with probability 1. This means that $\prod_{j=i+1}^{t} A_j \overset{\text{a.s.}}{\to} 0$ exponentially fast as $i \to -\infty$ and, hence, the right-hand infinite series in (4.2.28) converges a.s. (Verify this fact.) Write

$$Y_t' = 1 + \sum_{i=-\infty}^{t-1} \prod_{j=i+1}^{t} A_j = f(A_t, A_{t-1}, \ldots). \qquad (4.2.29)$$

For every fixed $n \ge 1$, the distribution of the vectors

$$\mathbf{A}_{t,n} = ((A_s)_{s \le t}, \ldots, (A_s)_{s \le t+n-1})$$

is independent of t, i.e., $\mathbf{A}_{t,n} \overset{d}{=} \mathbf{A}_{t+h,n}$ for every $t, h \in \mathbb{Z}$. Since f in (4.2.29) is a measurable function of $(A_s)_{s \le t}$, one may conclude that

$$(Y_t', \ldots, Y_{t+n-1}') \overset{d}{=} (Y_{t+h}', \ldots, Y_{t+h+n-1}') .$$

This means that (Y_t') is a *strictly stationary* sequence.[5] Obviously, (Y_t') is a solution to the stochastic recurrence equation (4.2.26). If there exists another strictly stationary sequence (Y_t'') satisfying (4.2.26), then iteration of (4.2.26) yields for $i \geq 1$,

$$|Y_t' - Y_t''| = A_t \cdots A_{t-i+1} \, |Y_{t-i}' - Y_{t-i}''| \,. \tag{4.2.30}$$

By the same argument as above,

$$A_t \cdots A_{t-i+1} = \exp\left\{ i \left[i^{-1} T_{t-i,t} \right] \right\} \stackrel{\text{a.s.}}{\to} 0$$

as $i \to \infty$, provided $E \log A_1 < 0$. Hence the right-hand side of (4.2.30) converges to zero in probability as $i \to \infty$ (verify this) and therefore $Y_t' = Y_t''$ a.s. Now we can identify the stationary sequence (Y_t') as the a.s. unique solution (Y_t) to the stochastic recurrence equation (4.2.26).

Since, by stationarity, $Y_t \stackrel{d}{=} Y_0$, it is not difficult to see that

$$Y_t \stackrel{d}{=} 1 + \sum_{i=-\infty}^{-1} \prod_{j=i+1}^{0} A_j \stackrel{d}{=} 1 + \sum_{i=1}^{\infty} \prod_{j=1}^{i} A_j \,.$$

Then we may conclude that for $x > 0$,

$$P(Y_0 > x)$$

$$\geq P\left(\sup_{n \geq 1} \prod_{j=1}^{n} A_j > x \right) = P\left(\sup_{n \geq 1} \sum_{j=1}^{n} \log A_j > \log x \right)$$

$$= \widetilde{\psi}(\log x) \,.$$

The event on the right-hand side reminds one of the skeleton process representation of the ruin event; see (4.1.2). Indeed, since $E \log A_1 < 0$ the process $S_n' = \sum_{j=1}^{n} \log A_j$ constitutes a random walk with negative drift as in the case of the ruin probability for the renewal model with NPC; see Section 4.1. If we interpret the random walk (S_n') as the skeleton process underlying a certain risk process, i.e., if we write $\log A_t = Z_t$, we can apply the bounds for the "ruin probability" $\widetilde{\psi}(x)$. For example, the Lundberg inequality yields

$$\widetilde{\psi}(\log x) \leq \exp\left\{ -r \log x \right\} = x^{-r}, \quad x \geq 1 \,,$$

provided that the equation $E A_1^h = E e^{h \log A_1} = 1$ has a unique positive solution r. The proof of this fact is analogous to the proof of Theorem 4.2.3.

This upper bound for $\widetilde{\psi}(\log x)$ does, however, not give one information about the decay of the tail $P(Y_0 > x)$. The Cramér bound of Theorem 4.2.5 is

---

[5] We refer to Brockwell and Davis [16] or Billingsley [13] for more information about stationary sequences.

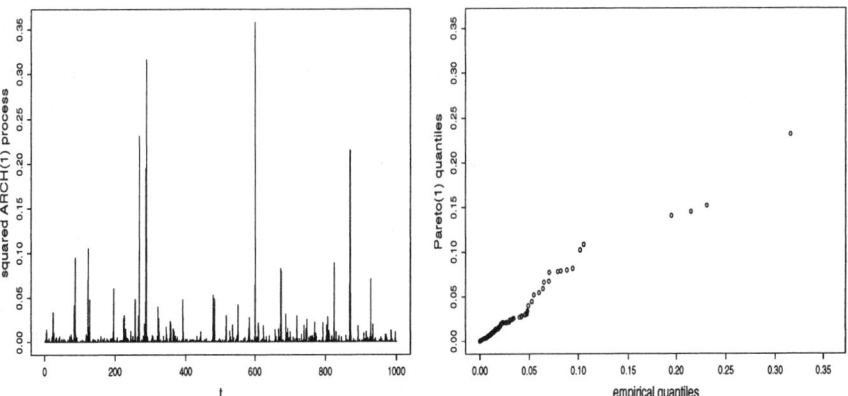

Figure 4.2.11 Left: *Simulation of* 1 000 *values* X_t^2 *from the squared* ARCH(1) *stochastic recurrence equation* (4.2.27) *with parameters* $\alpha_0 = 0.001$ *and* $\alpha_1 = 1$. *Since* var$(Z_1) = 1$ *the equation* $EA_1^h = E|Z_1|^{2h} = 1$ *has the unique positive solution* $r = 1$. *Thus we may conclude that* $P(X_t^2 > x) = C\,x^{-1}\,(1 + o(1))$ *for some positive constant* $C > 0$ *as* $x \to \infty$. *Right: QQ-plot of the sample of the squares* X_t^2 *against the Pareto distribution with tail parameter* 1. *The QQ-plot is in good agreement with the fact that the right tail of* X_1^2 *is Pareto like.*

in general not applicable since we required the Cramér-Lundberg model, i.e., we assumed that the quantities Z_t have the special structure $Z_t = X_t - cW_t$, where (W_t) is an iid exponential sequence, independent of the iid sequence (X_i). Nevertheless, it can be shown under additional conditions that the Cramér bound remains valid in this case, i.e., there exists a constant $C > 0$ such that

$$\widetilde{\psi}(\log x) = (1 + o(1))\,C\,e^{-r\,\log x} = (1 + o(1))\,C\,x^{-r}, \quad x \to \infty.$$

This gives a lower asymptotic power law bound for the tail $P(Y_0 > x)$. It can even be shown that this bound is precise:

$$P(Y_0 > x) = (1 + o(1))\,C\,x^{-r}, \quad x \to \infty,$$

provided that the "adjustment coefficient" $r > 0$ solves the equation $EA_1^h = 1$ and some further conditions on the distribution of A_1 are satisfied. We refer to Section 8.4 in Embrechts et al. [29] for an introduction to the subject of stochastic recurrence equations and related topics. The proofs in [29] are essentially based on work by Goldie [34]. Kesten [49] extended the results on power law tails for solutions to stochastic recurrence equations to the multivariate case. Power law tail behavior (regular variation) is a useful fact when one is interested in the analysis of extreme values in financial time series; see Mikosch [58] for a survey paper.   □

4.2.3 The Representation of the Ruin Probability as a Compound Geometric Probability

In this section we assume the Cramér-Lundberg model with NPC and use the notation of Section 4.2.2. Recall from Lemma 4.2.6 and (4.2.19) that the following equation for the non-ruin probability $\varphi = 1 - \psi$ was crucial for the derivation of Cramér's fundamental result:

$$\varphi(u) = \frac{\rho}{1+\rho} + \frac{1}{1+\rho} \int_0^u \varphi(u-y) \, dF_{X_1,I}(y) . \qquad (4.2.31)$$

According to the conditions in Lemma 4.2.6, for the validity of this equation one only needs to require that the claim sizes X_i have a density with finite expectation and that the NPC holds.

In this section we study equation (4.2.31) in some detail. First, we interpret the right-hand side of (4.2.31) as the distribution function of a compound geometric sum. Recall the latter notion from Example 3.3.2. Given a geometrically distributed random variable M,

$$p_n = P(M = n) = p \, q^n , \quad n = 0, 1, 2, \dots , \quad \text{for some } p = 1 - q \in (0, 1),$$

the random sum

$$S_M = \sum_{i=1}^M X_i$$

has a compound geometric distribution, provided M and the iid sequence (X_i) are independent. Straightforward calculation yields the distribution function

$$P(S_M \leq x) = p_0 + \sum_{n=1}^\infty p_n \, P(X_1 + \cdots + X_n \leq x)$$

$$= p + p \sum_{n=1}^\infty q^n \, P(X_1 + \cdots + X_n \leq x) . \qquad (4.2.32)$$

This result should be compared with the following one. In order to formulate it, we introduce a useful class of functions:

$$\mathcal{G} = \{G : \text{The function } G : \mathbb{R} \to [0, \infty) \text{ is non-decreasing, bounded,}$$

$$\text{right-continuous, and } G(x) = 0 \text{ for } x < 0\} .$$

In words, $G \in \mathcal{G}$ if and only if $G(x) = 0$ for negative x and there exist $c \geq 0$ and a distribution function F of a non-negative random variable such that $G(x) = c \, F(x)$ for $x \geq 0$.

Proposition 4.2.12 (Representation of the non-ruin probability as compound geometric probability)
Assume the Cramér-Lundberg model with $EX_1 < \infty$ and NPC. In addition, assume the claim sizes X_i have a density. Let $(X_{I,n})$ be an iid sequence with common distribution function $F_{X_1,I}$. Then the function φ given by

$$\varphi(u) = \frac{\rho}{1+\rho}\left[1 + \sum_{n=1}^{\infty}(1+\rho)^{-n} P(X_{I,1} + \cdots + X_{I,n} \leq u)\right], \quad u > 0.$$

$$(4.2.33)$$

satisfies (4.2.31). Moreover, the function φ defined in (4.2.33) is the only solution to (4.2.31) in the class $\mathcal{G}$.

The identity (4.2.33) will turn out to be useful since one can evaluate the right-hand side in some special cases. Moreover, a glance at (4.2.32) shows that the non-ruin probability φ has interpretation as the distribution function of a compound geometric sum with iid summands $X_{I,i}$ and $q = (1+\rho)^{-1}$.

Proof. We start by showing that φ given by (4.2.33) satisfies (4.2.31). It will be convenient to write $q = (1+\rho)^{-1}$ and $p = 1 - q = \rho(1+\rho)^{-1}$. Then we have

$$\varphi(u) = p + qp\Big[F_{X_1,I}(u) +$$

$$\sum_{n=2}^{\infty} q^{n-1} \int_0^u P(y + X_{I,2} + \cdots + X_{I,n} \leq u)\,dF_{X_1,I}(y)\Big]$$

$$= p + q\int_0^u p\left[1 + \sum_{n=1}^{\infty} q^n\, P(X_{I,1} + \cdots + X_{I,n} \leq u - y)\right]dF_{X_1,I}(y)$$

$$= p + q\int_0^u \varphi(u - y)\,dF_{X_1,I}(y)\,.$$

Hence φ satisfies (4.2.31).

It is not obvious that (4.2.33) is the only solution to (4.2.31) in the class $\mathcal{G}$. In order to show this it is convenient to use Laplace-Stieltjes transforms. The *Laplace-Stieltjes transform*[6] of a function $G \in \mathcal{G}$ is given by

$$\widehat{g}(t) = \int_{[0,\infty)} e^{-tx}\,dG(x)\,, \quad t \geq 0\,.$$

Notice that, for a distribution function G, $\widehat{g}(t) = Ee^{-tX}$, where X is a non-negative random variable with distribution function G. An important property

---

[6] The reader who would like to learn more about Laplace-Stieltjes transforms is referred for example to the monographs Bingham et al. [14], Feller [32] or Resnick [65]. See also Exercise 5 on p. 182 for some properties of Laplace-Stieltjes transforms.

of Laplace-Stieltjes transforms is that for any G_1, $G_2 \in \mathcal{G}$ with Laplace-Stieltjes transforms $\widehat{g}_1$, $\widehat{g}_2$, respectively, $\widehat{g}_1 = \widehat{g}_2$ implies that $G_1 = G_2$. This property can be used to show that φ given in (4.2.33) is the only solution to (4.2.31) in the class $\mathcal{G}$. We leave this as an exercise; see Exercise 5 on p. 182 for a detailed explanation of this problem. □

It is now an easy exercise to calculate $\psi(u)$ for exponential claim sizes by using Proposition 4.2.12.

Example 4.2.13 (The ruin probability in the Cramér-Lundberg model with exponential claim sizes)
For iid $\text{Exp}(\gamma)$ claim sizes X_i, Proposition 4.2.12 allows one to get an exact formula for $\psi(u)$. Indeed, formula (4.2.33) can be evaluated since the integrated tail distribution $F_{X_1,I}$ is again $\text{Exp}(\gamma)$ distributed and $X_{I,1} + \cdots + X_{I,n}$ has a $\Gamma(n,\gamma)$ distribution whose density is well-known. Use this information to prove that

$$\psi(u) = \frac{1}{1+\rho} \exp\left\{-\gamma \frac{\rho}{1+\rho} u\right\}, \quad u > 0.$$

Compare with Lundberg's inequality (4.2.12) in the case of exponential claim sizes. The latter bound is almost exact up to the constant multiple $(1+\rho)^{-1}$. □

4.2.4 Exact Asymptotics for the Ruin Probability: the Large Claim Case

In this section we again work under the hypothesis of the Cramér-Lundberg model with NPC.

The Cramér bound for the ruin probability $\psi(u)$

$$\psi(u) = Ce^{-ru}(1 + o(1)), \quad u \to \infty, \tag{4.2.34}$$

see Theorem 4.2.5, was obtained under a small claim condition: the existence of the moment generating function of X_1 in a neighborhood of the origin was a necessary assumption for the existence of the adjustment coefficient r given as the unique positive solution r to the equation $m_{Z_1}(h) = 1$.

It is the aim of this section to study what happens when the claim sizes are large. We learned in Section 3.2.6 that the subexponential distributions provide appropriate models of large claim sizes. The following result due to Embrechts and Veraverbeke [30] gives an answer to the ruin problem for large claims.

Theorem 4.2.14 (Ruin probability when the integrated claim size distribution is subexponential)
Assume the Cramér-Lundberg model with $EX_1 < \infty$ and NPC. In addition, assume that the claim sizes X_i have a density and that the integrated claim size distribution $F_{X_1,I}$ is subexponential. Then the ruin probability $\psi(u)$ satisfies the asymptotic relationship

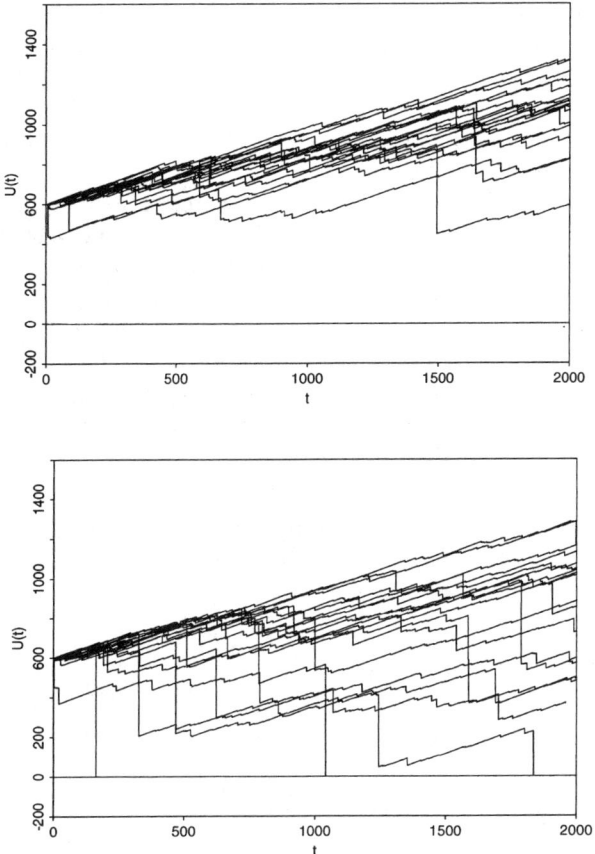

Figure 4.2.15 *Some realizations of the risk process U for log-normal (top) and Pareto distributed claim sizes* (bottom). *In the bottom graph one can see that ruin occurs due to a single very large claim size. This is typical for subexponential claim sizes.*

$$\lim_{u \to \infty} \frac{\psi(u)}{\overline{F}_{X_1,I}(u)} = \rho^{-1}. \qquad (4.2.35)$$

Embrechts and Veraverbeke [30] even showed the much stronger result that (4.2.35) is equivalent to each of the conditions $F_{X_1,I} \in \mathcal{S}$ and $(1 - \psi) \in \mathcal{S}$.

Relations (4.2.35) and the Cramér bound (4.2.34) show the crucial difference between heavy- and light-tailed claim size distributions. Indeed, (4.2.35) indicates that the probability of ruin $\psi(u)$ is essentially of the same order as $\overline{F}_{X_1,I}(u)$, which is non-negligible even if the initial capital u is large. For example, if the claim sizes are Pareto distributed with index $\alpha > 1$ (only in this case $EX_1 < \infty$), $\overline{F}_{X_1,I}$ is regularly varying with index $\alpha - 1$, and therefore

$\psi(u)$ decays at a power rate instead of an exponential rate in the light-tailed case. This means that portfolios with heavy-tailed claim sizes are dangerous; the largest claims have a significant influence on the overall behavior of the portfolio in a long term horizon. In contrast to the light-tailed claim size case, ruin happens spontaneously in the heavy-tailed case and is caused by one very large claim size; see Embrechts et al. [29], Section 8.3, for a theoretical explanation of this phenomenon.

The assumption of $F_{X_1,I}$ instead of F_{X_1} being subexponential is not verified in a straightforward manner even in the case of simple distribution functions F_{X_1} such as the log-normal or the Weibull ($\tau < 1$) distributions. There exists one simple case where one can verify subexponentiality of $F_{X_1,I}$ directly: the case of regularly varying F_{X_1} with index $\alpha > 1$. Then $F_{X_1,I}$ is regularly varying with index $\alpha - 1$; see Exercise 11 on p. 185. Sufficient conditions for $F_{X_1,I}$ to be subexponential are given in Embrechts et al. [29], p. 55. In particular, all large claim distributions collected in Table 3.2.19 are subexponential and so are their integrated tail distributions.

We continue with the proof of Theorem 4.2.14.

Proof. The key is the representation of the non-ruin probability $\varphi = 1 - \psi$ as compound geometric distribution, see Proposition 4.2.12, which in terms of ψ reads as follows:

$$\frac{\psi(u)}{\overline{F}_{X_1,I}(u)} = \frac{\rho}{1+\rho} \sum_{n=1}^{\infty} (1+\rho)^{-n} \frac{P(X_{I,1} + \cdots + X_{I,n} > u)}{\overline{F}_{X_1,I}(u)}.$$

By subexponentiality of $F_{X_1,I}$,

$$\lim_{u \to \infty} \frac{P(X_{I,1} + \cdots + X_{I,n} > u)}{\overline{F}_{X_1,I}(u)} = n, \quad n \ge 1.$$

Therefore a formal interchange of the limit $u \to \infty$ and the infinite series $\sum_{n=1}^{\infty}$ yields the desired relation:

$$\lim_{u \to \infty} \frac{\psi(u)}{\overline{F}_{X_1,I}(u)} = \frac{\rho}{1+\rho} \sum_{n=1}^{\infty} (1+\rho)^{-n} n = \rho^{-1}.$$

The justification of the interchange of limit and infinite series follows along the lines of the proof in Example 3.3.13 by using Lebesgue dominated convergence and exploiting the properties of subexponential distributions. We leave this verification to the reader. $\qquad \square$

Comments

The literature about ruin probabilities is vast. We refer to the monographs by Asmussen [4], Embrechts et al. [29], Grandell [36], Rolski et al. [67] for some recent overviews and to the literature cited therein. The notion of ruin probability can be directly interpreted in terms of the tail of the distribution of the stationary workload in a stable queue and therefore this notion also describes the average behavior of real-life queuing systems and stochastic networks.

The probability of ruin gives one a fine description of the long-run behavior in a homogeneous portfolio. In contrast to the results in Section 3.3, where the total claim amount $S(t)$ is treated as a *random variable* for fixed t or as $t \to \infty$, the ruin probability characterizes the total claim amount S as a *stochastic process*, i.e., as a random element assuming functions as values. The distribution of $S(t)$ for a fixed t is not sufficient for characterizing a complex quantity such as $\psi(u)$, which depends on the sample path behavior of S, i.e., on the whole distribution of the stochastic process.

The results of Cramér and Embrechts-Veraverbeke are of totally different nature; they nicely show the *phase transition* from heavy- to light-tailed distributions we have encountered earlier when we introduced the notion of subexponential distribution. The complete Embrechts-Veraverbeke result (Theorem 4.2.14 and its converse) shows that subexponential distributions constitute the most appropriate class of heavy-tailed distributions in the context of ruin. In fact, Theorem 4.2.14 can be dedicated to various authors; we refer to Asmussen [4], p. 260, for a historical account.

The ruin probability $\psi(u) = P(\inf_{t\geq0} U(t) < 0)$ is perhaps not the most appropriate risk measure from a practical point of view. Indeed, ruin in an infinite horizon is not the primary issue which an insurance business will actually be concerned about. As a matter of fact, ruin in a finite time horizon has also been considered in the above mentioned references, but it leads to more technical problems and often to less attractive theoretical results.

With a few exceptions, the ruin probability $\psi(u)$ cannot be expressed as an explicit function of the ingredients of the risk process. This calls for numerical or Monte Carlo approximations to $\psi(u)$, which is an even more complicated task than the approximation to the total claim amount distribution at a fixed instant of time. In particular, the subexponential case is a rather subtle issue. We again refer to the above-mentioned literature, in particular Asmussen [4] and Rolski et al. [67], who give overviews of the techniques needed.

Exercises

Sections 4.2.1 and 4.2.2
(1) Consider the Cramér-Lundberg model with Poisson intensity λ and $\Gamma(\gamma, \beta)$ distributed claim sizes X_i with density $f(x) = (\beta^\gamma / \Gamma(\gamma)) x^{\gamma-1} e^{-\beta x}$, $x > 0$.
(a) Calculate the moment generating function $m_{X_1}(h)$ of X_1. For which $h \in \mathbb{R}$ is the function well-defined?
(b) Derive the NPC.
(c) Calculate the adjustment coefficient under the NPC.
(d) Assume the claim sizes are $\Gamma(n, \beta)$ distributed for some integer $n \geq 1$. Write $\psi^{(n)}(u)$ for the corresponding ruin probability with initial capital $u > 0$. Suppose that the same premium $p(t) = c\,t$ is charged for $\Gamma(n, \beta)$ and $\Gamma(n+1, \beta)$ distributed claim sizes. Show that $\psi^{(n)}(u) \leq \psi^{(n+1)}(u)$, $u > 0$.
(2) Consider the risk process $U(t) = u + ct - S(t)$ in the Cramér-Lundberg model.

(a) Show that $S(s) = \sum_{i=1}^{N(s)} X_i$ is independent of $S(t) - S(s)$ for $s < t$. Hint: Use characteristic functions.

(b) Use (a) to calculate

$$E\left(e^{-hU(t)} \mid S(s)\right) \tag{4.2.36}$$

for $s < t$ and some $h > 0$. Here we assume that $Ee^{hS(t)}$ is finite. Under the assumption that the Lundberg coefficient r exists show the following relation:[7]

$$E\left(e^{-rU(t)} \mid S(s)\right) = e^{-rU(s)} \quad \text{a.s.} \tag{4.2.37}$$

(c) Under the assumptions of (b) show that $Ee^{-rU(t)}$ does not depend on t.

(3) Consider the risk process with premium rate c in the Cramér-Lundberg model with Poisson intensity λ. Assume that the adjustment coefficient r exists as the unique solution to the equation $1 = Ee^{r(X_1 - cW_1)}$. Write $m_A(t)$ for the moment generating function of any random variable A and $\rho = c/(\lambda EX_1) - 1 > 0$ for the safety loading. Show that r can be determined as the solution to each of the following equations.

$$\lambda + cr = \lambda m_{X_1}(r),$$

$$0 = \int_0^\infty [e^{rx} - (1+\rho)]\, P(X_1 > x)\, dx,$$

$$e^{cr} = m_{S(1)}(r),$$

$$c = \frac{1}{r} \log m_{S(1)}(r).$$

(4) Assume the Cramér-Lundberg model with the NPC. We also suppose that the moment generating function $m_{X_1}(h) = E\exp\{hX_1\}$ of the claim sizes X_i is finite for all $h > 0$. Show that there exists a unique solution $r > 0$ (Lundberg coefficient) to the equation $1 = E\exp\{h(X_1 - cW_1)\}$.

Section 4.2.3

(5) Let $\mathcal{G}$ be the class of non-decreasing, right-continuous, bounded functions G : $\mathbb{R} \to [0, \infty)$ such that $G(x) = 0$ for $x < 0$. Every such G can be written as $G = cF$ for some (probability) distribution function F of a non-negative random variable and some non-negative constant c. In particular, if $c = 1$, G is a distribution function. The *Laplace-Stieltjes transform* of $G \in \mathcal{G}$ is given by

$$\hat{g}(t) = \int_{[0,\infty)} e^{-tx}\, dG(x), \quad t \geq 0.$$

It is not difficult to see that $\hat{g}$ is well-defined. Here are some of the important properties of Laplace-Stieltjes transforms.

---

[7] The knowledgeable reader will recognize that (4.2.37) ensures that the process $M(t) = \exp\{-rU(t)\}$, $t \geq 0$, is a martingale with respect to the natural filtration generated by S, where one also uses the Markov property of S, i.e., $E(\exp\{-hU(t)\} \mid S(y), y \leq s) = E(\exp\{-hU(t)\} \mid S(s))$, $s < t$. Since the expectation of a martingale does not depend on t, we have $EM(t) = EM(0)$. This is the content of part (c) of this exercise.

(i) Different Laplace-Stieltjes transforms $\widehat{g}$ correspond to different functions $G \in \mathcal{G}$. This means the following: if $\widehat{g}_1$ is the Laplace-Stieltjes transform of $G_1 \in \mathcal{G}$ and $\widehat{g}_2$ the Laplace-Stieltjes transform of $G_2 \in \mathcal{G}$, then $\widehat{g}_1 = \widehat{g}_2$ implies that $G_1 = G_2$. See Feller [32], Theorem XIII.1.

(ii) Let $G_1, G_2 \in \mathcal{G}$ and $\widehat{g}_1, \widehat{g}_2$ be the corresponding Laplace-Stieltjes transforms. Write

$$(G_1 * G_2)(x) = \int_0^x G_1(x - y) \, dG_2(y) \,, \quad x \geq 0 \,,$$

for the *convolution* of G_1 and G_2. Then $G_1 * G_2$ has Laplace-Stieltjes transform $\widehat{g}_1 \, \widehat{g}_2$.

(iii) Let G^{n*} be the n-fold convolution of $G \in \mathcal{G}$, i.e., $G^{1*} = G$ and $G^{n*} = G^{(n-1)*} * G$. Then G^{n*} has Laplace-Stieltjes transform $\widehat{g}^n$.

(iv) The function $G = I_{[0,\infty)}$ has Laplace-Stieltjes transform $\widehat{g}(t) = 1$, $t \geq 0$.

(v) If $c \geq 0$ and $G \in \mathcal{G}$, cG has Laplace-Stieltjes transform $c\widehat{g}$.

(a) Show property (ii). Hint: Use the fact that for independent random variables A_1, A_2 with distribution functions G_1, G_2, respectively, the relation $(G_1 * G_2)(x) = P(A_1 + A_2 \leq x)$, $x \geq 0$, holds.

(b) Show properties (iii)-(v).

(c) Let H be a distribution function with support on $[0, \infty)$ and $q \in (0,1)$. Show that the function

$$G(u) = (1 - q) \sum_{n=0}^{\infty} q^n H^{n*}(u) \,, \quad u \geq 0 \,, \tag{4.2.38}$$

is a distribution function on $[0, \infty)$. We interpret $H^{0*} = I_{[0,\infty)}$.

(d) Let H be a distribution function with support on $[0, \infty)$ and with density h. Let $q \in (0,1)$. Show that the equation

$$G(u) = (1 - q) + q \int_0^u G(u - x) \, h(x) \, dx \,, \quad u \geq 0 \,. \tag{4.2.39}$$

has a solution G which is a distribution function with support on $[0, \infty)$. Hint: Look at the proof of Proposition 4.2.12.

(e) Show that (4.2.38) and (4.2.39) define the same distribution function G. Hint: Show that (4.2.38) and (4.2.39) have the same Laplace-Stieltjes transforms.

(f) Determine the distribution function G for $H \sim \text{Exp}(\gamma)$ by direct calculation from (4.2.38). Hint: H^{n*} is a $\Gamma(n, \gamma)$ distribution function.

(6) Consider the Cramér-Lundberg model with NPC, safety loading $\rho > 0$ and iid $\text{Exp}(\gamma)$ claim sizes.

(a) Show that the ruin probability is given by

$$\psi(u) = \frac{1}{1 + \rho} e^{-\gamma u \rho/(1+\rho)} \,, \quad u > 0 \,. \tag{4.2.40}$$

Hint: Use Exercise 5(f) and Proposition 4.2.12.

(b) Compare (4.2.40) with the Lundberg inequality.

(7) Consider the risk process $U(t) = u + ct - S(t)$ with total claim amount $S(t) = \sum_{i=1}^{N(t)} X_i$, where the iid sequence (X_i) of $\text{Exp}(\gamma)$ distributed claim sizes

is independent of the mixed homogeneous Poisson process N. In particular, we assume

$$(N(t))_{t\geq 0} = (\tilde{N}(\theta t))_{t\geq 0}\,,$$

where $\tilde{N}$ is a standard homogeneous Poisson process, independent of the positive mixing variable θ.

(a) Conditionally on θ, determine the NPC and the probability of ruin for this model, i.e.,

$$P\left(\inf_{t\geq 0} U(t) < 0 \,\Big|\, \theta\right).$$

(b) Apply the results of part (a) to determine the ruin probability

$$\psi(u) = P\left(\inf_{t\geq 0} U(t) < 0\right).$$

(c) Use part (b) to give conditions under which $\psi(u)$ decays exponentially fast to zero as $u \to \infty$.

(d) What changes in the above calculations if you choose the premium $p(t) = (1 + \rho)(\theta/\gamma)t$ for some $\rho > 0$? This means that you consider the risk process $U(t) = u + p(t) - S(t)$ with random premium adjusted to θ.

(8) Consider a reinsurance company with risk process $U(t) = u + ct - S(t)$, where the total claim amount $S(t) = \sum_{i=1}^{N(t)}(X_i - x)_+$ corresponds to an excess-of-loss treaty, see p. 148. Moreover, N is homogeneous Poisson with intensity λ, independent of the iid sequence (X_i) of $\mathrm{Exp}(\gamma)$ random variables. We choose the premium rate according to the expected value principle:

$$c = (1 + \rho)\,\lambda\,E[(X_1 - x)_+]$$

for some positive safety loading ρ.

(a) Show that $c = (1 + \rho)\,\lambda\,e^{-\gamma x}\,\gamma$.

(b) Show that

$$\phi_{(X_1-x)_+}(t) = E\,e^{\,i\,t\,(X_1-x)_+} = 1 + \frac{it}{\gamma - it}\,e^{-x\,\gamma}\,, \quad t \in \mathbb{R}\,.$$

(c) Show that $S(t)$ has the same distribution as $\tilde{S}(t) = \sum_{i=1}^{\tilde{N}(t)} X_i$, where $\tilde{N}$ is a homogeneous Poisson process with intensity $\tilde{\lambda} = \lambda\,e^{-\gamma x}$, independent of (X_i).

(d) Show that the processes S and $\tilde{S}$ have the same finite-dimensional distributions. Hint: The compound Poisson processes S and $\tilde{S}$ have independent stationary increments. See Corollary 3.3.9. Use (c).

(e) Define the risk process $\tilde{U}(t) = u + ct - \tilde{S}(t)$, $t \geq 0$. Show that

$$\psi(u) = P\left(\inf_{t\geq 0} U(t) < 0\right) = P\left(\inf_{t\geq 0} \tilde{U}(t) < 0\right)$$

and calculate $\psi(u)$. Hint: Use (d).

Section 4.2.4

(9) Give a detailed proof of Theorem 4.2.14.

(10) Verify that the integrated tail distribution corresponding to a Pareto distribution is subexponential.

(11) Let $f(x) = x^\delta L(x)$ be a regularly varying function, where L is slowly varying and δ is a real number; see Definition 3.2.20. A well-known result which runs under the name *Karamata's theorem* (see Feller [32]) says that, for any $y_0 > 0$,

$$\lim_{y \to \infty} \frac{\int_y^\infty f(x)\, dx}{y\, f(y)} = -(1+\delta)^{-1} \text{ if } \delta < -1$$

and

$$\lim_{y \to \infty} \frac{\int_{y_0}^y f(x)\, dx}{y\, f(y)} = (1+\delta)^{-1} \text{ if } \delta > -1.$$

Use this result to show that the integrated tail distribution of any regularly varying distribution with index $\alpha > 1$ is subexponential.

Part II

Experience Rating

In Part I we focused on the overall or average behavior of a homogeneous insurance portfolio, where the claim number process occurred independently of the iid claim size sequence. As a matter of fact, this model disregards the policies, where the claims come from. For example, in a portfolio of car insurance policies the driving skill and experience, the age of the driver, the gender, the profession, etc., are factors which are not of interest. The policy-holders generate iid claims which are aggregated in the total claim amount. The goal of collective risk theory is to determine the order of magnitude of the total claim amount in order to judge the risk represented by the claims in the portfolio as time goes by.

Everybody will agree that it is to some extent unfair and perhaps even unwise if every policyholder had to pay the same premium. A driver with poor driving skills would have to pay the same premium as a policyholder who drives carefully and has never caused any accident in his/her life. Therefore it seems reasonable to build an *individual model* for every policyholder which takes his or her claim history into account for determining a premium, as well as the overall behavior of the portfolio. This is the basic idea of *credibility theory*, which was popularized and propagated by Hans Bühlmann in his monograph [19] and in the articles [17, 18]. The monograph [19] was one of the first rigorous treatments of non-life insurance which used modern probability theory. It is one of the classics in the field and has served generations of actuaries as a guide for insurance mathematics.

In Chapter 5 we sketch the theory on *Bayes estimation* of the premium for an individual policy based on the data available in the policy. Instead of the expected total claim amount, which was the crucial quantity for the premium calculation principles in a *portfolio* (see Section 3.1.3), premium calculation in a *policy* is based on the expected claim size/claim number, conditionally on the experience in the policy. This so-called *Bayes estimator* of the individual premium minimizes the mean square deviation from the conditional expectation in the class of all finite variance measurable functions of the data. Despite the elegance of the theory, the generality of the class of approximating functions leads to problems when it comes to determining the Bayes estimator for concrete examples.

For this reason, the class of *linear Bayes* or *credibility estimators* is introduced in Chapter 6. Here the mean square error is minimized over a subclass of all measurable functions of the data having finite variance: the class of *linear* functions of the data. This minimization procedure leads to mathematically tractable expressions. The coefficients of the resulting *linear Bayes estimator* are determined as the solution to a system of linear equations. It turns out that the linear Bayes estimator can be understood as the convex combination of the overall portfolio mean and of the sample mean in the individual policy. Depending on the experience in the policy, more or less weight is given to the individual experience or to the portfolio experience. This means that the data of the policy become *more credible* if a lot of experience about the policy is available. This is the fundamental idea of *credibility theory*. We consider the basics on linear Bayes estimation in Section 6.1. In Sections 6.2-6.4 we apply the theory to two of the best known models in this context: the Bühlmann and the Bühlmann-Straub models.

5

Bayes Estimation

In this chapter we consider the basics of experience rating in a policy. The heterogeneity model is fundamental. It combines the experience about the claims in an individual policy with the experience of the claims in the whole portfolio; see Section 5.1. In this model, a random parameter is attached to every policy. According to the outcome of this parameter in a particular policy, the distribution of the claims in the policy is chosen. This random *heterogeneity parameter* determines essential properties of the policy. Conditionally on this parameter, the expected claim size (or claim number) serves as a means for determining the premium in the policy. Since the heterogeneity parameter of a policy is not known a priori, one uses the data of the policy to *estimate* the conditional expectation in the policy. In this chapter, an estimator is obtained by minimizing the mean square deviation of the estimator (which can be any finite variance measurable function of the data) from the conditional expectation in the policy. The details of this so-called *Bayes estimation* procedure and the estimation error are discussed in Section 5.2. There we also give some intuition on the name *Bayes estimator*.

5.1 The Heterogeneity Model

In this section we introduce an *individual model* which describes one particular policy and its inter-relationship with the portfolio. We assume that the claim history of the ith policy in the portfolio is given by a time series of non-negative observations

$$x_{i,1}, \ldots, x_{i,n_i} .$$

The latter sequence of numbers is interpreted as a realization of the sequence of non-negative random variables

$$X_{i,1}, \ldots, X_{i,n_i} .$$

Here $X_{i,t}$ is interpreted as the claim size or the claim number occurring in the ith policy in the tth period. Periods can be measured in months, half-years, years, etc. The number n_i is then the sample size in the ith policy.

A natural question to ask is

How can one determine a premium for the ith policy by taking the claim history into account?

A simple means to determine the premium would be to calculate the expectation of the $X_{i,t}$'s. For example, if $(X_{i,t})_{t \geq 1}$ constituted an iid sequence and n_i were large we could use the strong law of large numbers to get an approximation of $EX_{i,t}$:

$$\overline{X}_i = \frac{1}{n_i} \sum_{t=1}^{n_i} X_{i,t} \approx EX_{i,1} \quad \text{a.s.}$$

There are, however, some arguments against this approach. If n_i is not large enough, the variation of $\overline{X}_i$ around the mean $EX_{i,1}$ can be quite large which can be seen by a large variance $\text{var}(\overline{X}_i)$, provided the latter quantity is finite. Moreover, if a new policy started, no experience about the policyholder would be available: $n_i = 0$. One can also argue that the claims caused in one policy are not really independent. For example, in car insurance the individual driver is certainly a factor which has significant influence on the size and the frequency of the claims.

Here an additional modeling idea is needed: to every policy we assign a random parameter θ which contains essential information about the policy. For example, it tells one how much driving skill or experience the policyholder has. Since one usually does not know these properties before the policy is purchased, one assumes that the sequence of θ_i's, where θ_i corresponds to the ith policy, constitutes an iid random sequence. This means that all policies behave on average in the same way; what matters is the random realization $\theta_i(\omega)$ which determines the individual properties of the ith policy, and the totality of the values θ_i determines the heterogeneity in the portfolio.

Definition 5.1.1 (The heterogeneity model)

(1) *The ith policy is described by the pair $(\theta_i, (X_{i,t})_{t \geq 1})$, where the random parameter θ_i is the* heterogeneity parameter *and $(X_{i,t})_{t \geq 1}$ is the sequence of claim sizes or claim numbers in the policy.*
(2) *The sequence of pairs $(\theta_i, (X_{i,t})_{t \geq 1})$, $i = 1, 2, \ldots$, is iid.*
(3) *Given θ_i, the sequence $(X_{i,t})_{t \geq 1}$ is iid with distribution function $F(\cdot | \theta_i)$.*

The conditions of this model imply that the claim history of the ith policy, given by the sequence of claim sizes or claim numbers, is mutually independent of the other policies. This is a natural condition which says that the different policies do not interfere with each other. Dependence is only possible between the claim sizes/claim numbers $X_{i,t}$, $t = 1, 2, \ldots$, within the ith portfolio. The assumption that these random variables are iid conditionally on θ_i is certainly an idealization which has been made for mathematical convenience. Later, in Chapter 6, we will replace this assumption by a weaker condition.

The $X_{i,t}$'s are identically distributed with distribution function

$$P(X_{i,t} \le x) = E[P(X_{i,t} \le x \mid \theta_i)] = E[P(X_{i,1} \le x \mid \theta_i)]$$

$$= E[F(x \mid \theta_i)] = E[F(x \mid \theta_1)].$$

Now we come back to the question how we could determine a premium in the ith policy by taking into account the individual claim history. Since expectations $EX_{i,t}$ are not sensible risk measures in this context, a natural surrogate quantity is given by

$$\mu(\theta_i) = E(X_{i,1} \mid \theta_i) = \int_{\mathbb{R}} x \, dF(x \mid \theta_i),$$

where we assume the latter quantity is well-defined, the condition $EX_{1,1} < \infty$ being sufficient. Notice that $\mu(\theta_i)$ is a measurable function of the random variable θ_i. Since the sequence (θ_i) is iid, so is $(\mu(\theta_i))$.

In a sense, $\mu(\theta_i)$ can be interpreted as a net premium (see Section 3.1.3) in the ith policy which gives one an idea how much premium one should charge.

Under the conditions of the heterogeneity model, the strong law of large numbers implies that $\overline{X}_i \overset{\text{a.s.}}{\to} \mu(\theta_i)$ as $n_i \to \infty$. (Verify this relation! Hint: first apply the strong law of large numbers conditionally on θ_i.) Therefore $\overline{X}_i$ can be considered as one possible approximation to $\mu(\theta_i)$. It is the aim of the next section to show how one can find *best approximations* (in the mean square sense) to $\mu(\theta_i)$ from the available data. These so-called *Bayes estimators* or not necessarily linear functions of the data.

5.2 Bayes Estimation in the Heterogeneity Model

In this section we assume the heterogeneity model; see Definition 5.1.1. It is our aim to find a reasonable approximation to the quantity $\mu(\theta_i) = E(X_{i,1} \mid \theta_i)$ by using all available data $X_{i,t}$.

Write

$$\mathbf{X}_i = (X_{i,1}, \dots, X_{i,n_i})', \quad i = 1, \dots, r,$$

for the samples of data available in the r independent policies. Since the samples are mutually independent, it seems unlikely that $\mathbf{X}_j$, $j \ne i$, will contain any useful information about $\mu(\theta_i)$. This conjecture will be confirmed soon.

In what follows, we assume that $\text{var}(\mu(\theta_i))$ is finite. Then it makes sense to consider the quantity

$$\rho(\widehat{\mu}) = E\left[(\mu(\theta_i) - \widehat{\mu})^2\right],$$

where $\widehat{\mu}$ is any measurable real-valued function of the data $\mathbf{X}_1, \ldots, \mathbf{X}_r$ with finite variance. The notation $\rho(\widehat{\mu})$ is slightly misleading since ρ is not a function of the random variable $\widehat{\mu}$ but of the joint distribution of $(\widehat{\mu}, \mu(\theta_i))$. We will nevertheless use this symbol since it is intuitively appealing.

We call the quantity $\rho(\widehat{\mu})$ the (quadratic) *risk* or the *mean square error* of $\widehat{\mu}$ (with respect to $\mu(\theta_i)$). The choice of the quadratic risk is mainly motivated by mathematical tractability.[1] We obtain an approximation (estimator) $\widehat{\mu}_B$ to $\mu(\theta_i)$ by minimizing $\rho(\widehat{\mu})$ over a suitable class of distributions of $(\mu(\theta_i), \widehat{\mu})$.

Theorem 5.2.1 (Minimum risk estimation of $\mu(\theta_i)$)
The minimizer of the risk $\rho(\widehat{\mu})$ in the class of all measurable functions $\widehat{\mu}$ of $\mathbf{X}_1, \ldots, \mathbf{X}_r$ with $\mathrm{var}(\widehat{\mu}) < \infty$ exists and is unique with probability 1. It is attained for

$$\widehat{\mu}_B = E(\mu(\theta_i) \mid \mathbf{X}_i)$$

with corresponding risk

$$\rho(\widehat{\mu}_B) = E[\mathrm{var}(\mu(\theta_i) \mid \mathbf{X}_i)].$$

The index B indicates that $\widehat{\mu}_B$ is a so-called *Bayes estimator*. We will give an argument for the choice of this name in Example 5.2.4 below.

Proof of Theorem 5.2.1. The result is a special case of a well-known fact on conditional expectations which we recall and prove here for convenience.

Lemma 5.2.2 *Let X be a random variable defined on the probability space $(\Omega, \mathcal{G}, P)$ and $\mathcal{F}$ be a sub-σ-field of $\mathcal{G}$. Assume $\mathrm{var}(X) < \infty$. Denote the set of random variables on $(\Omega, \mathcal{F}, P)$ with finite variance by $L^2(\Omega, \mathcal{F}, P)$. Then the minimizer of $E[(X-Y)^2]$ in the class of all random variables $Y \in L^2(\Omega, \mathcal{F}, P)$ exists and is a.s. unique. It is attained at $Y = E(X \mid \mathcal{F})$ with probability 1.[2]*

Proof. Since both X and Y have finite variance and live on the same probability space, we can define $E[(X - Y)^2]$ and $E(X \mid \mathcal{F})$. Then

$$E[(X - Y)^2] = E\left[([X - E(X \mid \mathcal{F})] + [E(X \mid \mathcal{F}) - Y])^2\right]. \quad (5.2.1)$$

Notice that $X - E(X \mid \mathcal{F})$ and $E(X \mid \mathcal{F}) - Y$ are uncorrelated. Indeed, $X - E(X \mid \mathcal{F})$ has mean zero, and exploiting the fact that both Y and $E(X \mid \mathcal{F})$ are $\mathcal{F}$-measurable,

---

[1] The theory in Chapters 5 and 6 is based on Hilbert space theory; the resulting estimators can be interpreted as projections from the space of all square integrable random variables into smaller Hilbert sub-spaces.
[2] If one wants to be mathematically correct, one has to consider $L^2(\Omega, \mathcal{F}, P)$ as the collection of equivalence classes of random variables modulo P whose representatives have finite variance and are $\mathcal{F}$-measurable.

$$E\Big([X - E(X \mid \mathcal{F})]\,[E(X \mid \mathcal{F}) - Y]\Big)$$

$$= E\Big(E\big[[X - E(X \mid \mathcal{F})]\,[E(X \mid \mathcal{F}) - Y]\,\big|\, \mathcal{F}\big]\Big)$$

$$= E\Big([E(X \mid \mathcal{F}) - Y]\,E[X - E(X \mid \mathcal{F}) \mid \mathcal{F}]\Big)$$

$$= E\Big([E(X \mid \mathcal{F}) - Y]\,[E(X \mid \mathcal{F}) - E(X \mid \mathcal{F})]\Big)$$

$$= 0\,.$$

Hence relation (5.2.1) becomes

$$E[(X - Y)^2] = E\left([X - E(X \mid \mathcal{F})]^2\right) + E\left([E(X \mid \mathcal{F}) - Y]^2\right)$$

$$\geq E\left([X - E(X \mid \mathcal{F})]^2\right)\,.$$

Obviously, in the latter inequality one achieves equality if and only if $Y = E(X \mid \mathcal{F})$ a.s. This means that minimization in the class $L^2(\Omega, \mathcal{F}, P)$ of all $\mathcal{F}$-measurable random variables Y with finite variance yields $E(X \mid \mathcal{F})$ as the only candidate, with probability 1. □

Now turn to the proof of the theorem. We denote by $\mathcal{F} = \sigma(\mathbf{X}_1, \ldots, \mathbf{X}_r)$ the sub-σ-field generated by the data $\mathbf{X}_1, \ldots, \mathbf{X}_r$. Then the theorem aims at minimizing

$$\rho(\widehat{\mu}) = E[(\mu(\theta_i) - \widehat{\mu})^2]$$

in the class $L^2(\Omega, \mathcal{F}, P)$ of finite variance measurable functions $\widehat{\mu}$ of the data $\mathbf{X}_1, \ldots, \mathbf{X}_r$. This is the same as saying that $\widehat{\mu}$ is $\mathcal{F}$-measurable and $\mathrm{var}(\widehat{\mu}) < \infty$. Then Lemma 5.2.2 tells us that the minimizer of $\rho(\widehat{\mu})$ exists, is a.s. unique and given by

$$\widehat{\mu}_{\mathrm{B}} = E(\mu(\theta_i) \mid \mathcal{F}) = E(\mu(\theta_i) \mid \mathbf{X}_1, \ldots, \mathbf{X}_r) = E(\mu(\theta_i) \mid \mathbf{X}_i)\,.$$

In the last step we used the fact that θ_i and $\mathbf{X}_j$, $j \neq i$, are mutually independent.

It remains to calculate the risk:

$$\rho(\widehat{\mu}_{\mathrm{B}}) = E\left[(\mu(\theta_i) - E(\mu(\theta_i) \mid \mathbf{X}_i))^2\right]$$

$$= E\left(E\left[(\mu(\theta_i) - E(\mu(\theta_i) \mid \mathbf{X}_i))^2 \,\big|\, \mathbf{X}_i\right]\right)$$

$$= E[\mathrm{var}(\mu(\theta_i) \mid \mathbf{X}_i)]\,.$$

This proves the theorem. □

From Theorem 5.2.1 it is immediate that the minimum risk estimator $\widehat{\mu}_{\mathrm{B}}$ only depends on the data in the ith portfolio. Therefore we suppress the index i

in the notation wherever we focus on one particular policy. We write θ for θ_i and $X_1, X_2, \ldots$ for $X_{i,1}, X_{i,2}, \ldots$, but also $\mathbf{X}$ instead of $\mathbf{X}_i$ and n instead of n_i.

The calculation of the Bayes estimator $E(\mu(\theta) \mid \mathbf{X})$ very much depends on the knowledge of the conditional distribution of $\theta \mid \mathbf{X}$. The following lemma contains some useful rules how one can calculate the conditional density $\theta \mid \mathbf{X}$ provided the latter exists.

Lemma 5.2.3 (Calculation of the conditional density of θ given the data) *Assume the heterogeneity model, that θ has density f_θ and the conditional density $f_\theta(y \mid \mathbf{X} = \mathbf{x})$, $y \in \mathbb{R}$, of the one-dimensional parameter θ given $\mathbf{X}$ exists for $\mathbf{x}$ in the support of $\mathbf{X}$.*

(1) *If X_1 has a discrete distribution then $\theta \mid \mathbf{X}$ has density*

$$f_\theta(y \mid \mathbf{X} = \mathbf{x}) \tag{5.2.2}$$

$$= \frac{f_\theta(y) \, P(X_1 = x_1 \mid \theta = y) \cdots P(X_1 = x_n \mid \theta = y)}{P(\mathbf{X} = \mathbf{x})}, \quad y \in \mathbb{R},$$

on the support of $\mathbf{X}$.

(2) *If $(\mathbf{X}, \theta)$ have the joint density $f_{\mathbf{X},\theta}$, then $\theta \mid \mathbf{X}$ has density*

$$f_\theta(y \mid \mathbf{X} = \mathbf{x}) = \frac{f_\theta(y) \, f_{X_1}(x_1 \mid \theta = y) \cdots f_{X_1}(x_n \mid \theta = y)}{f_{\mathbf{X}}(\mathbf{x})}, \quad y \in \mathbb{R},$$

on the support of $\mathbf{X}$.

Proof. (1) Since the conditional density of $\theta \mid \mathbf{X}$ is assumed to exist we have

$$P(\theta \le x \mid \mathbf{X} = \mathbf{x}) = \int_{-\infty}^{x} f_\theta(y \mid \mathbf{X} = \mathbf{x}) \, dy, \quad x \in \mathbb{R}. \tag{5.2.3}$$

Since the X_i's are iid conditionally on θ, for $x \in \mathbb{R}$,

$$P(\theta \le x \mid \mathbf{X} = \mathbf{x})$$

$$= [P(\mathbf{X} = \mathbf{x})]^{-1} \, E[P(\theta \le x, \mathbf{X} = \mathbf{x} \mid \theta)]$$

$$= [P(\mathbf{X} = \mathbf{x})]^{-1} \, E[I_{(-\infty,x]}(\theta) \, P(\mathbf{X} = \mathbf{x} \mid \theta)]$$

$$= [P(\mathbf{X} = \mathbf{x})]^{-1} \int_{-\infty}^{x} P(\mathbf{X} = \mathbf{x} \mid \theta = y) \, f_\theta(y) \, dy$$

$$= \int_{-\infty}^{x} [P(\mathbf{X} = \mathbf{x})]^{-1} \, P(X_1 = x_1 \mid \theta = y) \cdots P(X_1 = x_1 \mid \theta = y) \, f_\theta(y) \, dy.$$

$$\tag{5.2.4}$$

By the Radon-Nikodym theorem, the integrands in (5.2.3) and (5.2.4) coincide a.e. This gives (5.2.2).

(2) The conditional density of $\mathbf{X} \mid \theta$ satisfies

$$f_{\mathbf{X}}(\mathbf{x} \mid \theta = y) = f_{\mathbf{X},\theta}(\mathbf{x}, y)/f_\theta(y),$$

on the support of θ, see for example Williams [78], Section 15.6. On the other hand, in the heterogeneity model the $\mathbf{X}_i$'s are iid given θ. Hence

$$f_{\mathbf{X}}(\mathbf{x} \mid \theta) = f_{X_1}(x_1 \mid \theta) \cdots f_{X_1}(x_n \mid \theta).$$

We conclude that

$$f_\theta(y \mid \mathbf{X} = \mathbf{x}) = \frac{f_{\theta,\mathbf{X}}(y, \mathbf{x})}{f_{\mathbf{X}}(\mathbf{x})} = \frac{f_\theta(y)\, f_{X_1}(x_1 \mid \theta = y) \cdots f_{X_1}(x_n \mid \theta = y)}{f_{\mathbf{X}}(\mathbf{x})}.$$

This concludes the proof of (2).   □

Example 5.2.4 (Poisson distributed claim numbers and gamma distributed heterogeneity parameters)
Assume the claim numbers X_t, $t = 1, 2, \ldots$, are iid with $\mathrm{Pois}(\theta)$ distribution, given θ, and $\theta \sim \Gamma(\gamma, \beta)$ for some positive γ and β, i.e.,

$$f_\theta(x) = \frac{\beta^\gamma}{\Gamma(\gamma)} x^{\gamma-1} e^{-\beta x}, \quad x > 0.$$

It was mentioned in Example 2.3.3 that X_t is then negative binomially distributed with parameter $(\beta/(1 + \beta), \gamma)$. Also recall that

$$E\theta = \frac{\gamma}{\beta} \quad \text{and} \quad \mathrm{var}(\theta) = \frac{\gamma}{\beta^2}. \tag{5.2.5}$$

Since $X_1 \mid \theta$ is $\mathrm{Pois}(\theta)$ distributed,

$$\mu(\theta) = E(X_1 \mid \theta) = \theta.$$

We intend to calculate the Bayes estimator $\widehat{\mu}_{\mathrm{B}} = E(\theta \mid \mathbf{X})$ of θ. We start by calculating the distribution of θ given $\mathbf{X}$. We apply formula (5.2.2):

$$f_\theta(x \mid \mathbf{X} = \mathbf{x})$$

$$= P(X_1 = x_1 \mid \theta = x) \cdots P(X_n = x_n \mid \theta = x)\, f_\theta(x)\, [P(\mathbf{X} = \mathbf{x})]^{-1}$$

$$= D_1(\mathbf{x})\, x^{\gamma-1} e^{-\beta x} \prod_{t=1}^n \left(\frac{x^{x_t}}{x_t!} e^{-x} \right)$$

$$= D_2(\mathbf{x})\, x^{\gamma + x. -1} e^{-x(\beta + n)}, \tag{5.2.6}$$

where $D_1(\mathbf{x})$ and $D_2(\mathbf{x})$ are certain multipliers which do not depend on x, and $x. = \sum_{t=1}^n x_t$. Since (5.2.6) represents a density, we may conclude from its particular form that it is the density of the $\Gamma(\gamma + x., \beta + n)$ distribution, i.e., $\theta \mid \mathbf{X} = \mathbf{x}$ has this particular gamma distribution.

From (5.2.5) we can deduce the expectation and variance of $\theta \mid \mathbf{X}$:

$$E(\theta \mid \mathbf{X}) = \frac{\gamma + X.}{\beta + n} \quad \text{and} \quad \text{var}(\theta \mid \mathbf{X}) = \frac{\gamma + X.}{(\beta + n)^2},$$

where $X. = \sum_{t=1}^n X_t$. Hence the Bayes estimator $\widehat{\mu}_{\mathrm{B}}$ of $\mu(\theta) = \theta$ is

$$\widehat{\mu}_{\mathrm{B}} = \frac{\gamma + X.}{\beta + n}$$

and the corresponding risk is given by

$$\rho(\widehat{\mu}_{\mathrm{B}}) = E(\text{var}(\theta \mid \mathbf{X})) = E\left(\frac{\gamma + X.}{(\beta + n)^2}\right) = \frac{\gamma + n\, EX_1}{(\beta + n)^2} = \frac{\gamma}{\beta}\frac{1}{\beta + n},$$

where we used the fact that $EX_1 = E[E(X_1 \mid \theta)] = E\theta = \gamma/\beta$.
The Bayes estimator $\widehat{\mu}_{\mathrm{B}}$ of θ has representation

$$\widehat{\mu}_{\mathrm{B}} = (1 - w)\, E\theta + w\, \overline{X},$$

where $\overline{X} = n^{-1}X.$ is the sample mean in the policy and

$$w = \frac{n}{\beta + n}$$

is a positive weight. Thus the Bayes estimator of θ given the data $\mathbf{X}$ is a weighted mean of the expected heterogeneity parameter $E\theta$ and the sample mean in the individual policy. Notice that $w \to 1$ if the sample size $n \to \infty$. This means that the Bayes estimator $\widehat{\mu}_{\mathrm{B}}$ gets closer to $\overline{X}$ the larger the sample size. For small n, the variation of $\overline{X}$ is too large in order to be representative of the policy. Therefore the weight w given to the policy average $\overline{X}$ is small, whereas the weight $1 - w$ assigned to the expected value $E\theta$ of the portfolio heterogeneity is close to one. This means that the net premium represented by $\mu(\theta) = E(X_1 \mid \theta) = \theta$ is strongly influenced by the information available in the policy. In particular, if no such information is available, i.e., $n = 0$, premium calculation is solely based on the overall portfolio expectation. Also notice that the risk satisfies

$$\rho(\widehat{\mu}_{\mathrm{B}}) = (1 - w)\, \text{var}(\theta) = (1 - w)\frac{\gamma}{\beta^2} \to 0 \quad \text{as } n \to \infty.$$

Finally, we comment on the name *Bayes estimator*. It stems from Bayesian statistics, which forms a major part of modern statistics. Bayesian statistics has gained a lot of popularity over the years, in particular, since Bayesian techniques have taken advantage of modern computer power. One of the fundamental ideas of this theory is that the parameter of a distribution is not deterministic but has distribution in the parameter space considered. In the context of our example, we *assumed* that the parameter θ has a gamma distribution with given parameters γ and β. This distribution has to be known (conjectured) in advance and is therefore referred to as the *prior distribution*. Taking into account the information which is represented by the sample **X**, we then updated the distribution of θ, i.e., we were able to calculate the distribution of $\theta \mid \mathbf{X}$ and obtained the gamma distribution with parameters $\gamma + X$. and $\beta + n$. We see from this example that the data change the prior distribution in a particular way. The resulting gamma distribution is referred to as the *posterior distribution*. This reasoning might explain the notion of Bayes estimator. □

Comments

The minimization of the risk $\rho(\widehat{\mu})$ in the class of all finite variance measurable functions of the data leads in general to a situation where one cannot calculate the Bayes estimator $\widehat{\mu}_B = E(\mu(\theta) \mid \mathbf{X})$ explicitly. In the next section we will therefore minimize the risk over the smaller class of linear functions of the data and we will see that this estimator can be calculated explicitly.

The idea of minimizing over the class of all measurable functions is basic to various concepts in probability theory and statistics. In this section we have already seen that the conditional expectation of a random variable with respect to a σ-field is such a concept. Similar concepts occur in the context of predicting future values of a time series based on the information contained in the past, in regression analysis, Kalman filtering or extrapolation in spatial processes. As a matter of fact, we have calculated an approximation to the "best prediction" $\mu(\theta_i) = E(X_{i,n_i+1} \mid \theta_i)$ of the next claim size/number X_{i,n_i+1} in the ith policy by minimizing the quadratic risk $E[(E(X_{i,n_i+1} \mid \theta_i) - \widehat{\mu})^2]$ in the class of all measurable functions of the data $X_{i,1}, \ldots, X_{i,n_i}$. Therefore the idea underlying the Bayes estimator considered in this section has been exploited in other areas as well and the theory in these other fields is often directly interpretable in terms of Bayes estimation. We refer for example to Brockwell and Davis [16] for prediction of time series and Kalman filtering, and to Cressie's book [24] on spatial statistics.

Parts of standard textbooks on statistics are devoted to Bayesian statistics. We refer to the classical textbook of Lehmann [53] for an introduction to the theory. Bühlmann's monograph [17] propagated the use of Bayesian methods for premium calculation in a policy. Since then, major parts of textbooks on non-life insurance mathematics have been devoted to the Bayes methodology; see for example Kaas et al. [46], Klugman et al. [51], Sundt [77], Straub [75].

Exercises

(1) Assume the heterogeneity model.
(a) Give a necessary and sufficient condition for the independence of $X_{i,t}$, $t = 1, \ldots, n_i$, in the ith policy.
(b) Assume that $EX_{1,1} < \infty$. Show that $E(X_{i,1} \mid \theta_i)$ is well-defined and finite. Prove the following strong laws of large numbers as $n \to \infty$:

$$\frac{1}{n}\sum_{t=1}^{n} X_{i,t} \overset{\text{a.s.}}{\to} \mu(\theta_i) = E(X_{i,1} \mid \theta_i) \quad \text{and} \quad \frac{1}{n}\sum_{i=1}^{n} X_{i,t} \overset{\text{a.s.}}{\to} EX_{1,1}.$$

(2) Assume the heterogeneity model and consider the ith policy. We suppress the dependence on i in the notation. Given $\theta > 0$, let the claim sizes $X_1, \ldots, X_n$ in the policy be iid Pareto distributed with parameters (λ, θ), i.e.,

$$\overline{F}(x \mid \theta) = P(X_i > x \mid \theta) = (\lambda/x)^{\theta}, \quad x > \lambda.$$

Assume that θ is $\Gamma(\gamma, \beta)$ distributed with density

$$f_{\gamma,\beta}(x) = \frac{\beta^{\gamma}}{\Gamma(\gamma)} x^{\gamma-1} e^{-\beta x}, \quad x > 0.$$

(a) Show that $\theta \mid \mathbf{X}$ with $\mathbf{X} = (X_1, \ldots, X_n)'$ has density

$$f_{\gamma+n,\beta+\sum_{i=1}^{n}\log(X_i/\lambda)}(x).$$

(b) A reinsurance company takes into account only the values X_i exceeding a known high threshold K. They "observe" the counting variables $Y_i = I_{(K,\infty)}(X_i)$ for a known threshold $K > \lambda$. The company is interested in estimating $P(X_1 > K \mid \theta)$.

(i) Give a naive estimator of $P(X_1 > K \mid \theta)$ based on the empirical distribution function of $X_1, \ldots, X_n$.

(ii) Determine the a.s. limit of this estimator as $n \to \infty$. Does it coincide with $P(X_1 > K \mid \theta)$?

(c) Show that Y_i, given θ, is $\text{Bin}(1, p(\theta))$ distributed, where $p(\theta) = E(Y_1 \mid \theta)$. Compare $p(\theta)$ with the limit in (b,ii).

(d) Show that the Bayes estimator of $p(\theta) = E(Y_1 \mid \theta)$ based on the data $Y_1, \ldots, Y_n$ is given by

$$\frac{(\beta + \sum_{i=1}^{n}\log(X_i/\lambda))^{\gamma+n}}{(\beta + \sum_{i=1}^{n}\log(X_i/\lambda) + \log(K/\lambda))^{\gamma+n}}.$$

(3) Assume the heterogeneity model and consider a policy with one observed claim number X and corresponding heterogeneity parameter θ. We assume that $X \mid \theta$ is $\text{Pois}(\theta)$ distributed, where θ has a continuous density f_θ on $(0, \infty)$. Notice that $E(X \mid \theta) = \theta$.

(a) Determine the conditional density $f_\theta(y \mid X = k)$, $k = 0, 1, \ldots$, of $\theta \mid X$ and use this information to calculate the Bayes estimator $m_k = E(\theta \mid X = k)$, $k = 0, 1, 2, \ldots$.

(b) Show that

$$m_k = (k+1)\frac{P(X = k+1)}{P(X = k)}, \quad k = 0, 1, \ldots.$$

(c) Show that

$$E(\theta^l \mid X = k) = \prod_{i=0}^{l-1} m_{k+i}, \quad k \geq 0, \, l \geq 1.$$

(4) Consider the ith policy in a heterogeneity model. We suppress the dependence on i in the notation. We assume the heterogeneity parameter θ to be $\beta(a,b)$-distributed with density

$$f_\theta(y) = \frac{\Gamma(a+b)}{\Gamma(a)\,\Gamma(b)}\, y^{a-1}\,(1-y)^{b-1}, \quad 0 < y < 1, \quad a,b > 0.$$

Given θ, the claim numbers $X_1, \ldots, X_n$ are iid $\mathrm{Bin}(k,\theta)$ distributed.

(a) Calculate the conditional density $f_\theta(y \mid \mathbf{X} = \mathbf{x})$ of θ given $\mathbf{X} = (X_1, \ldots, X_n)' = \mathbf{x} = (x_1, \ldots, x_n)'$.

(b) Calculate the Bayes estimator $\widehat{\mu}_B$ of $\mu(\theta) = E(X_1 \mid \theta)$ and the corresponding risk. Hint: A $\beta(a,b)$-distributed random variable θ satisfies the relations $E\theta = a/(a+b)$ and $\mathrm{var}(\theta) = ab/[(a+b+1)(a+b)^2]$.

(5) Consider the ith policy in a heterogeneity model. We suppress the dependence on i in the notation. We assume the heterogeneity parameter θ to be $N(\mu, \sigma^2)$-distributed. Given θ, the claim sizes $X_1, \ldots, X_n$ are iid log-normal (θ, τ)-distributed. This means that $\log X_t$ has representation $\log X_t = \theta + \tau Z_t$ for an iid $N(0,1)$ sequence (Z_t) independent of θ and some positive constant τ.

(a) Calculate the conditional density $f_\theta(y \mid \mathbf{X} = \mathbf{x})$ of θ given $\mathbf{X} = (X_1, \ldots, X_n)' = \mathbf{x} = (x_1, \ldots, x_n)'$.

(b) Calculate the Bayes estimator $\widehat{\mu}_B$ of $\mu(\theta) = E(X_1 \mid \theta)$ and the corresponding risk. It is useful to remember that

$$E e^{a+b\,Z_1} = e^{a+b^2/2} \quad \text{and} \quad \mathrm{var}\left(e^{a+b\,Z_1}\right) = e^{2a+b^2}\left(e^{b^2} - 1\right), \quad a \in \mathbb{R}, b > 0.$$

6

Linear Bayes Estimation

As mentioned at the end of Chapter 5, it is generally difficult, if not impossible, to calculate the Bayes estimator $\widehat{\mu}_{\mathrm{B}} = E(\mu(\theta_i) \mid \mathbf{X}_i)$ of the net premium $\mu(\theta_i) = E(X_{i,t} \mid \theta)$ in the ith policy based on the data $\mathbf{X}_i = (X_{i,1}, \ldots, X_{i,n_i})'$. As before, we write $X_{i,t}$ for the claim size/claim number in the ith policy in the tth period. One way out of this situation is to minimize the risk,

$$\rho(\widehat{\mu}) = E\left[(\mu(\theta_i) - \widehat{\mu})^2 \right],$$

not over the whole class of finite variance measurable functions $\widehat{\mu}$ of the data $\mathbf{X}_1, \ldots, \mathbf{X}_r$, but over a smaller class. In this section we focus on the class of linear functions

$$\mathcal{L} = \left\{ \widehat{\mu} : \widehat{\mu} = a_0 + \sum_{i=1}^{r} \sum_{t=1}^{n_i} a_{i,t} X_{i,t}, \quad a_0, a_{i,t} \in \mathbb{R} \right\}. \qquad (6.0.1)$$

If a minimizer of the risk $\rho(\widehat{\mu})$ in the class $\mathcal{L}$ exists, we call it a *linear Bayes estimator* for $\mu(\theta_i)$, and we denote it by $\widehat{\mu}_{\mathrm{LB}}$.

We start in Section 6.1 by solving the above minimization problem in a wider context: we consider the best approximation (with respect to quadratic risk) of a finite variance random variable by linear functions of a given vector of finite variance random variables. The coefficients of the resulting linear function and the corresponding risk can be expressed as the solution to a system of linear equations, the so-called *normal equations*. This is an advantage compared to the Bayes estimator, where, in general, we could not give an explicit solution to the minimization problem. In Section 6.2 we apply the minimization result to the original question about estimation of the conditional policy mean $\mu(\theta_i)$ by linear functions of the data $\mathbf{X}_1, \ldots, \mathbf{X}_n$. It turns out that the requirements of the heterogeneity model (Definition 5.1.1) can be relaxed. Indeed, the heterogeneity model is tailored for Bayes estimation, which requires one to specify the complete dependence structure inside and across the policies. Since linear Bayes estimation is concerned with the minimization of second moments, it is plausible in this context that one only needs to assume suitable conditions about the first and second moments inside and across the policies. These attempts result in the so-called *Bühlmann model* of Section 6.2 and, in a more general context, in the *Bühlmann-Straub model* of Section 6.4. In Sections 6.3 and 6.4 we also derive the corresponding linear Bayes estimators and their risks.

6.1 An Excursion to Minimum Linear Risk Estimation

In this section we consider the more general problem of approximating a finite variance random variable X by linear functions of finite variance random variables $Y_1, \ldots, Y_m$ which are defined on the same probability space. Write $\mathbf{Y} = (Y_1, \ldots, Y_m)'$. Then our task is to approximate X by any element of the class of linear functions

$$\mathcal{L}' = \{Y : Y = a_0 + \mathbf{a}'\mathbf{Y}, \quad a_0 \in \mathbb{R}, \mathbf{a} \in \mathbb{R}^m\}, \qquad (6.1.2)$$

where $\mathbf{a} = (a_1, \ldots, a_m)' \in \mathbb{R}^m$ is any column vector. In Section 6.3 we will return to the problem of estimating $X = \mu(\theta_i)$ by linear functions of the data $\mathbf{X}_1, \ldots, \mathbf{X}_r$. There we will apply the theory developed in this section.

We introduce the *expectation vector* of the vector $\mathbf{Y}$:

$$E\mathbf{Y} = (EY_1, \ldots, EY_m)',$$

the *covariance vector* of X and $\mathbf{Y}$:

$$\Sigma_{X,\mathbf{Y}} = (\mathrm{cov}(X, Y_1), \ldots, \mathrm{cov}(X, Y_m))'$$

and the *covariance matrix* of $\mathbf{Y}$:

$$\Sigma_{\mathbf{Y}} = (\mathrm{cov}(Y_i, Y_j))_{i,j=1,\ldots,m},$$

where we assume that all quantities are well-defined and finite.

The following auxiliary result gives a complete answer to the approximation problem of X in the class $\mathcal{L}'$ of linear functions Y of the random variables Y_i with respect to quadratic risk $E[(X - Y)^2]$.

Proposition 6.1.1 (Minimum risk estimation by linear functions)
Assume that $\mathrm{var}(X) < \infty$ *and* $\mathrm{var}(Y_i) < \infty$, $i = 1, \ldots, m$. *Then the following statements hold.*

(1) *Let* $(a_0, \mathbf{a})$ *be any solution of the system of linear equations*

$$a_0 = EX - \mathbf{a}' E\mathbf{Y}, \quad \Sigma'_{X,\mathbf{Y}} = \mathbf{a}' \Sigma_{\mathbf{Y}}, \qquad (6.1.3)$$

and $\widehat{Y} = a_0 + \mathbf{a}'\mathbf{Y}$. *Then for any* $Y \in \mathcal{L}'$ *the risk* $E[(X - Y)^2]$ *is bounded from below by*

$$E[(X - Y)^2] \geq E[(X - \widehat{Y})^2] = \text{var}(X) - \mathbf{a}' \, \Sigma_{\mathbf{Y}} \, \mathbf{a}, \qquad (6.1.4)$$

and the right-hand side does not depend on the particular choice of the solution $(a_0, \mathbf{a})$ to (6.1.3). This means that any $\widehat{Y} \in \mathcal{L}'$ with $(a_0, \mathbf{a})$ satisfying (6.1.3) is a minimizer of the risk $E[(X - Y)^2]$. Conversely, (6.1.3) is a necessary condition for $\widehat{Y}$ to be a minimizer of the risk.

(2) *The estimator $\widehat{Y}$ of X introduced in (1) satisfies the equations*

$$EX = E\widehat{Y}, \quad \text{cov}(X, Y_i) = \text{cov}(\widehat{Y}, Y_i), \quad i = 1, \ldots, m. \qquad (6.1.5)$$

(3) *If $\Sigma_{\mathbf{Y}}$ has inverse, then there exists a unique minimizer $\widehat{Y}$ of the risk $E[(X - Y)^2]$ in the class $\mathcal{L}'$ given by*

$$\widehat{Y} = EX + \Sigma'_{X,\mathbf{Y}} \, \Sigma_{\mathbf{Y}}^{-1} \, (\mathbf{Y} - E\mathbf{Y}). \qquad (6.1.6)$$

with risk given by

$$E[(X - \widehat{Y})^2] = \text{var}(X) - \Sigma'_{X,\mathbf{Y}} \, \Sigma_{\mathbf{Y}}^{-1} \, \Sigma_{X,\mathbf{Y}} \qquad (6.1.7)$$

$$= \text{var}(X) - \text{var}(\widehat{Y}). \qquad (6.1.8)$$

It is not difficult to see that (6.1.3) always has a solution $(a_0, \mathbf{a})$ (we have $m + 1$ linear equations for the $m + 1$ variables a_i), but it is not necessarily unique. However, any $\widehat{Y} = a_0 + \mathbf{a}' \, \mathbf{Y}$ with $(a_0, \mathbf{a})$ satisfying (6.1.3) has the same (minimal) risk.

Relations (6.1.7)-(6.1.8) imply that

$$\text{var}(\widehat{Y}) = \Sigma'_{X,\mathbf{Y}} \, \Sigma_{\mathbf{Y}}^{-1} \, \Sigma_{X,\mathbf{Y}} \, .$$

and that $\widehat{Y}$ and $X - \widehat{Y}$ are uncorrelated.

Proof. (1) We start by verifying necessary conditions for the existence of a minimizer $\widehat{Y}$ of the risk in the class $\mathcal{L}'$. In particular, we will show that (6.1.3) is a necessary condition for $\widehat{Y} = a_0 + \mathbf{a}'\mathbf{Y}$ to minimize the risk. Since the smallest risk $E[(X - Y)^2]$ for any $Y = a_0 + \mathbf{a}' \, \mathbf{Y} \in \mathcal{L}'$ can be written in the form

$$\inf_{\mathbf{a}, a_0} E\left[(X - (a_0 + \mathbf{a}' \, \mathbf{Y}))^2\right] = \inf_{\mathbf{a}} \inf_{a_0} E\left[(X - (a_0 + \mathbf{a}' \, \mathbf{Y}))^2\right],$$

one can use a two-step minimization procedure:

(a) Fix $\mathbf{a}$ and minimize the risk $E[(X - Y)^2]$ with respect to a_0.
(b) Plug the a_0 from (a) into the risk $E[(X - Y)^2]$ and minimize with respect to $\mathbf{a}$.

For fixed $\mathbf{a}$ and any $Y \in \mathcal{L}'$, $E[(X - Y)^2] \geq \text{var}(X - Y)$ since $E(Z + c)^2 \geq \text{var}(Z)$ for any random variable Z and any constant $c \in \mathbb{R}$. Therefore the first of the equations in (6.1.3) determines a_0. It ensures that $EX = E\widehat{Y}$. Since we fixed $\mathbf{a}$, the minimizer a_0 is a function of $\mathbf{a}$. Now plug this particular $\mathbf{a}$ into the risk. Then straightforward calculation yields:

$$E[(X - Y)^2] = \mathrm{var}(X - Y)$$

$$= E\left[\left((X - EX) - \sum_{t=1}^{m} a_t\,(Y_t - EY_t)\right)^2\right]$$

$$= \mathrm{var}(X) + \mathrm{var}\left(\sum_{t=1}^{m} a_t\,Y_t\right) - 2\,\mathrm{cov}\left(X\,,\sum_{t=1}^{m} a_t\,Y_t\right)$$

$$= \mathrm{var}(X) + \sum_{t=1}^{m}\sum_{s=1}^{m} a_t\,a_s\,\mathrm{cov}(Y_t, Y_s) - 2\sum_{t=1}^{m} a_t\,\mathrm{cov}(X, Y_t)\,. \qquad (6.1.9)$$

Differentiating the latter relation with respect to a_k and setting the derivatives equal to zero, one obtains the system of linear equations

$$0 = \sum_{t=1}^{m} a_t\,\mathrm{cov}(Y_k, Y_t) - \mathrm{cov}(X, Y_k)\,, \quad k = 1,\ldots, m\,.$$

Using the notation introduced at the beginning of this section, we see that the latter equation says nothing but

$$\Sigma'_{X,Y} = \mathbf{a}'\,\Sigma_Y\,, \qquad (6.1.10)$$

which is the desired second equation in (6.1.3).

So far we have proved that the coefficients $(a_0, \mathbf{a})$ of any minimizer $\hat{Y} = a_0 + \mathbf{a}'\mathbf{Y}$ of the risk $E[(X - Y)^2]$ in the class $\mathcal{L}'$ necessarily satisfy relation (6.1.3). To complete the proof it remains to show that any solution to (6.1.3) minimizes the risk $E[(X - Y)^2]$ in $\mathcal{L}'$. One way to show this is by considering the matrix of second partial derivatives of (6.1.9) as a function of $\mathbf{a}$. Direct calculation shows that this matrix is Σ_Y. Any covariance matrix is non-negative definite which condition is sufficient for the existence of a minimum of the function (6.1.9) at $\mathbf{a}$ satisfying the necessary condition (6.1.3). A unique minimizer exists if the matrix of second partial derivatives is positive definite. This condition is satisfied if and only if Σ_Y is invertible.

An alternative way to verify that any $\hat{Y}$ with $(a_0, \mathbf{a})$ satisfying (6.1.3) minimizes the risk goes as follows. Pick any $Y \in \mathcal{L}'$ with representation $Y = b_0 + \mathbf{b}'\mathbf{Y}$. Then

$$E[(X - Y)^2] \geq \mathrm{var}(X - Y) \qquad (6.1.11)$$

$$= E\left[\left([(X - EX) - \mathbf{a}'\,(\mathbf{Y} - E\mathbf{Y})] + (\mathbf{a} - \mathbf{b})'\,(\mathbf{Y} - E\mathbf{Y})\right)^2\right]\,.$$

Since the coefficients a_t satisfy relation (6.1.10) it is not difficult to verify that the random variables $X - \mathbf{a}'\,\mathbf{Y}$ and $(\mathbf{a} - \mathbf{b})'\,\mathbf{Y}$ are uncorrelated. Hence we conclude from (6.1.11) and (6.1.10) that

$$E[(X - Y)^2] \geq \mathrm{var}\,(X - \mathbf{a'\,Y}) + \mathrm{var}\,((\mathbf{a} - \mathbf{b})'\mathbf{Y})$$

$$\geq \mathrm{var}\,(X - \mathbf{a'\,Y})$$

$$= \mathrm{var}(X) + \mathrm{var}(\mathbf{a'Y}) - 2\,\mathrm{cov}(X, \mathbf{a'\,Y})$$

$$= \mathrm{var}(X) + \mathbf{a'}\,\Sigma_{\mathbf{Y}}\,\mathbf{a} - 2\,\mathbf{a'}\,\Sigma_{X,\mathbf{Y}}$$

$$= \mathrm{var}(X) - \mathbf{a'}\,\Sigma_{\mathbf{Y}}\,\mathbf{a}\,.$$

This relation implies that for any $Y \in \mathcal{L}'$ the risk $E[(X - Y)^2]$ is bounded from below by the risk $E[(X - (a_0 + \mathbf{a'\,Y}))^2]$ for any $(a_0, \mathbf{a})$ satisfying (6.1.3). It remains to show that the risk does not depend on the particular choice of $(a_0, \mathbf{a})$. Suppose both $\widehat{Y}, \widetilde{Y} \in \mathcal{L}'$ have coefficients satisfying (6.1.3). But then $E[(X - \widehat{Y})^2] \geq E[(X - \widetilde{Y})^2] \geq E[(X - \widehat{Y})^2]$. Hence they have the same risk.

(2) We have to show the equivalence of (6.1.3) and (6.1.5). If (6.1.3) holds,

$$\widehat{Y} = a_0 + \mathbf{a'\,Y} = EX + \mathbf{a'}\,(\mathbf{Y} - E\mathbf{Y})\,,$$

and hence the identity $E\widehat{Y} = EX$ is obvious. If (6.1.5) holds, take expectations in $\widehat{Y} = a_0 + \mathbf{a'\,Y}$ to conclude that $a_0 = EX - \mathbf{a'}E\mathbf{Y}$.

It is straightforward to see that

$$\mathrm{cov}(\widehat{Y}, Y_i) = \mathrm{cov}(\mathbf{a'Y}, Y_i) = \mathbf{a'}\,\Sigma_{Y_i, \mathbf{Y}}\,, \quad i = 1, \ldots, m\,. \qquad (6.1.12)$$

Assuming (6.1.3), the latter relations translate into

$$\Sigma'_{\widehat{Y}, \mathbf{Y}} = \mathbf{a'}\,\Sigma_{\mathbf{Y}} = \Sigma'_{X, \mathbf{Y}}\,.$$

This proves the equality of the covariances in (6.1.5). Conversely, assuming (6.1.5) and again exploiting (6.1.12), it is straightforward to see that

$$\mathrm{cov}(X, Y_i) = \mathbf{a'}\,\Sigma_{Y_i, \mathbf{Y}}\,, \quad i = 1, \ldots, m\,,$$

implying the second relation in (6.1.3).

(3) From the first equation of (6.1.3) we know that any minimizer $\widehat{Y}$ of the risk in $\mathcal{L}'$ can be written in the form

$$\widehat{Y} = a_0 + \sum_{t=1}^{m} a_t\,Y_t = [EX - \mathbf{a'}\,E\mathbf{Y}] + \mathbf{a'Y} = EX + \mathbf{a'}\,(\mathbf{Y} - E\mathbf{Y})\,.$$

Moreover, the system of linear equations $\Sigma'_{X, \mathbf{Y}} = \mathbf{a'}\,\Sigma_{\mathbf{Y}}$ in (6.1.3) has a unique solution if and only if $\Sigma_{\mathbf{Y}}^{-1}$ exists, and then

$$\Sigma'_{X\,\mathbf{Y}}\,\Sigma_{\mathbf{Y}}^{-1} = \mathbf{a'}\,.$$

Plugging the latter relation into $\widehat{Y}$, we obtain

$$\widehat{Y} = EX + \Sigma'_{X,Y} \Sigma_Y^{-1} (\mathbf{Y} - E\mathbf{Y}).$$

This is the desired relation (6.1.6) for $\widehat{Y}$. The risk is derived in a similar way by taking into account the right-hand side of relation (6.1.4). This proves (6.1.7). Relation (6.1.8) follows by observing that $\operatorname{var}(\widehat{Y}) = \operatorname{var}(\mathbf{a}'\mathbf{Y}) = \mathbf{a}'\Sigma_{\mathbf{Y}}\mathbf{a}$.   □

Both relations (6.1.3) and (6.1.5) determine the minimum risk estimator $\widehat{Y}$ of X in the class $\mathcal{L}'$ of linear functions of the Y_t's. Because of their importance they get a special name.

Definition 6.1.2 (Normal equations, linear Bayes estimator)
Each of the equivalent relations (6.1.3) and (6.1.5) is called the normal equations. *The minimum risk estimator* $\widehat{Y} = a_0 + \mathbf{a}'\,\mathbf{Y}$ *in the class* $\mathcal{L}'$ *of linear functions of the* Y_t*'s, which is determined by the normal equations, is the* linear Bayes estimator *of* X.

The name "linear Bayes estimator" is perhaps not most intuitive in this general context. We choose it because linear Bayes estimation will be applied to $X = \mu(\theta_i)$ in the next sections, where we want to compare it with the more complex Bayes estimator of $\mu(\theta_i)$ introduced in Chapter 5.

6.2 The Bühlmann Model

Now we return to our original problem of determining the minimum risk estimator of $\mu(\theta_i)$ in the class $\mathcal{L}$, see (6.0.1). An analysis of the proof of Proposition 6.1.1 shows that only expectations, variances and covariances were needed to determine the linear Bayes estimator. For this particular reason we introduce a model which is less restrictive than the general heterogeneity model; see Definition 5.1.1. The following model fixes the conditions for linear Bayes estimation.

Definition 6.2.1 (The Bühlmann model)

(1) *The ith policy is described by the pair* $(\theta_i, (X_{i,t})_{t \geq 1})$, *where the random parameter* θ_i *is the* heterogeneity parameter *and* $(X_{i,t})_{t \geq 1}$ *is the sequence of claim sizes or claim numbers in the policy.*
(2) *The pairs* $(\theta_i, (X_{i,t})_{t \geq 1})$ *are mutually independent.*
(3) *The sequence* (θ_i) *is iid.*
(4) *Conditionally on* θ_i, *the* $X_{i,t}$*'s are independent and their expectation and variance are given functions of* θ_i:

$$\mu(\theta_i) = E(X_{i,t} \mid \theta_i) \quad and \quad v(\theta_i) = \operatorname{var}(X_{i,t} \mid \theta_i).$$

Since the functions $\mu(\theta_i)$ and $v(\theta_i)$ only depend on θ_i, it follows that $(\mu(\theta_i))$ and $(v(\theta_i))$ are iid sequences. It will be convenient to use the following notation:

$$\mu = E\mu(\theta_i), \quad \lambda = \mathrm{var}(\mu(\theta_i)) \quad \text{and} \quad \varphi = Ev(\theta_i).$$

The Bühlmann model differs from the heterogeneity model in the following aspects:

- The sequence $((X_{i,t})_{t\geq 1})_{i\geq 1}$ consists of independent components $(X_{i,t})_{t\geq 1}$ which are not necessarily identically distributed.
- In particular, the $X_{i,t}$'s inside and across the policies can have different distributions.
- Only the conditional expectation $\mu(\theta_i)$ and the conditional variance $v(\theta_i)$ are the same for $X_{i,t}$, $t = 1, 2, \ldots$. The remaining distributional characteristics of the $X_{i,t}$'s are not fixed.

The heterogeneity model is a special case of the Bühlmann model insofar that in the former case the random variables $X_{i,t}$, $t = 1, 2, \ldots$, are iid given θ_i and that the $X_{i,t}$'s are identically distributed for all i, t.

We mention that the first two moments of the $X_{i,t}$'s are the same for all i and t, and so are the covariances. Since we will make use of these facts quite often, we collect here some of the relations needed.

Lemma 6.2.2 *Assume the conditions of the Bühlmann model and that the variances $\mathrm{var}(X_{i,t})$ are finite for all i and t. Then the following relations are satisfied for $i \geq 1$ and $t \neq s$:*

$$EX_{i,t} = E[E(X_{i,t} \mid \theta_i)] = E\mu(\theta_i) = \mu,$$

$$E(X_{i,t}^2) = E[E(X_{i,t}^2 \mid \theta_i)] = E[\mathrm{var}(X_{i,t} \mid \theta_i)] + E[(E(X_{i,t} \mid \theta_i))^2]$$

$$= \varphi + E[(\mu(\theta_i))^2] = \varphi + \lambda + \mu^2,$$

$$\mathrm{var}(X_{i,t}) = \varphi + \lambda,$$

$$\mathrm{cov}(X_{i,t}, X_{i,s}) = E[E(X_{i,t} - EX_{i,1} \mid \theta_i)\,E(X_{i,s} - EX_{i,1} \mid \theta_i)]$$

$$= \mathrm{var}(\mu(\theta_i)) = \lambda,$$

$$\mathrm{cov}(\mu(\theta_i), X_{i,t}) = E[(\mu(\theta_i) - EX_{i,1})\,E[X_{i,t} - EX_{i,1} \mid \theta_i]] = \mathrm{var}(\mu(\theta_i)) = \lambda.$$

Remark 6.2.3 By virtue of Lemma 6.2.2, the covariance matrix $\Sigma_{\mathbf{X}_i}$ is rather simple:

$$\mathrm{cov}(X_{i,t}, X_{i,s}) = \begin{cases} \lambda + \varphi & \text{if } t = s, \\ \lambda & \text{if } t \neq s. \end{cases}$$

Therefore the inverse of $\Sigma_{\mathbf{X}_i}$ exists if and only if $\varphi > 0$, i.e., $\mathrm{var}(X_{i,t} \mid \theta_i)$ is not equal to zero a.s. This is a very natural condition. Indeed, if $\varphi = 0$ one has $X_{i,t} = \mu(\theta_i)$ a.s., i.e., there is no variation inside the policies.

6.3 Linear Bayes Estimation in the Bühlmann Model

Writing

$$\mathbf{Y} = \text{vec}(\mathbf{X}_1, \ldots, \mathbf{X}_r) = (X_{1,1}, \ldots, X_{1,n_1}, \ldots, X_{r,1}, \ldots, X_{r,n_r})',$$

$$\mathbf{a} = \text{vec}(\mathbf{a}_1, \ldots, \mathbf{a}_r) = (a_{1,1}, \ldots, a_{1,n_1}, \ldots, a_{r,1}, \ldots, a_{r,n_r})',$$

we can identify $\mathcal{L}$ in (6.0.1) and $\mathcal{L}'$ in (6.1.2). Then Proposition 6.1.1 applies.

Theorem 6.3.1 (Linear Bayes estimator in the Bühlmann model)
Consider the Bühlmann model. Assume $\text{var}(X_{i,t}) < \infty$ *for all* i, t *and* $\varphi > 0$.
Then the linear Bayes estimator $\widehat{\mu}_{\text{LB}} = a_0 + \mathbf{a}'\mathbf{Y}$ *of* $\mu(\theta_i) = E(X_{i,t} \mid \theta_i)$ *in the class* $\mathcal{L}$ *of the linear functions of the data* $\mathbf{X}_1, \ldots, \mathbf{X}_r$ *exists, is unique and given by*

$$\widehat{\mu}_{\text{LB}} = (1-w)\,\mu + w\,\overline{X}_i, \tag{6.3.13}$$

where

$$w = \frac{n_i\,\lambda}{\varphi + n_i\,\lambda}. \tag{6.3.14}$$

The risk of $\widehat{\mu}_{\text{LB}}$ *is given by*

$$\rho(\widehat{\mu}_{\text{LB}}) = (1-w)\,\lambda.$$

Similarly to the Bayes estimator $\widehat{\mu}_{\text{B}}$ we observe that $\widehat{\mu}_{\text{LB}}$ only depends on the data $\mathbf{X}_i$ of the ith policy. This is not surprising in view of the independence of the policies.

It is worthwhile comparing the linear Bayes estimator (6.3.13) with the Bayes estimator in the special case of Example 5.2.4. Both are weighted means of $EX_{i,t} = \mu$ and $\overline{X}_i$. In general, the Bayes estimator does not have such a linear representation; see for example Exercise 2 on p. 200.

Proof. We have to verify the normal equations (6.1.3) for $X = \mu(\theta_i)$ and $\mathbf{Y}$ as above. Since the policies are independent, $X_{i,t}$ and $X_{j,s}$, $i \neq j$, are independent. Hence

$$\text{cov}(X_{i,t}, X_{j,s}) = 0 \quad \text{for } i \neq j \text{ and any } s, t.$$

Therefore the second equation in (6.1.3) turns into

$$\mathbf{0} = \mathbf{a}_j'\,\Sigma_{\mathbf{X}_j}, \quad j \neq i, \quad \Sigma_{\mu(\theta_i),\mathbf{X}_i}' = \mathbf{a}_i'\,\Sigma_{\mathbf{X}_i}.$$

For $j \neq i$, $\mathbf{a}_j = \mathbf{0}$ is the only possible solution since $\Sigma_{\mathbf{X}_j}^{-1}$ exists; see Remark 6.2.3. Therefore the second equation in (6.1.3) turns into

$$\Sigma_{\mu(\theta_i),\mathbf{X}_i}' = \mathbf{a}_i'\,\Sigma_{\mathbf{X}_i}, \quad \mathbf{a}_j = \mathbf{0}, \quad j \neq i. \tag{6.3.15}$$

Since $EX_{i,t} = \mu$ and also $E\mu(\theta_i) = \mu$, see Lemma 6.2.2, the first equation in (6.1.3) yields

$$a_0 = \mu \left(1 - a_{i,\cdot}\right), \qquad (6.3.16)$$

where $a_{i,\cdot} = \sum_{t=1}^{n_i} a_{i,t}$. Relations (6.3.15) and (6.3.16) imply that the linear Bayes estimator of $\mu(\theta_i)$ only depends on the data $\mathbf{X}_i$ of the ith policy. For this reason, we suppress the index i in the notation for the rest of the proof.

An appeal to (6.3.15) and Lemma 6.2.2 yields

$$\lambda = a_t \operatorname{var}(X_1) + (a_{\cdot} - a_t) \operatorname{var}(\mu(\theta)) = a_t (\lambda + \varphi) + (a_{\cdot} - a_t) \lambda$$

$$= a_t \varphi + a_{\cdot} \lambda, \quad t = 1, \ldots, n. \qquad (6.3.17)$$

This means that $a_t = a_1$, $t = 1, \ldots, n$, with

$$a_1 = \frac{\lambda}{\varphi + n\lambda}.$$

Then, by (6.3.16),

$$a_0 = \mu \left(1 - n\, a_1\right) = \mu \frac{\varphi}{\varphi + n\lambda}.$$

Finally, write $w = n\, a_1$. Then

$$\widehat{\mu}_{\mathrm{LB}} = a_0 + \mathbf{a}' \mathbf{X} = (1 - w)\, \mu + a_1\, X_{\cdot} = (1 - w)\, \mu + w\, \overline{X}.$$

Now we are left to derive the risk of $\widehat{\mu}_{\mathrm{LB}}$. From (6.1.8) and Lemma 6.2.2 we know that

$$\rho(\widehat{\mu}_{\mathrm{LB}}) = \operatorname{var}(\mu(\theta)) - \operatorname{var}(\widehat{\mu}_{\mathrm{LB}}) = \lambda - \operatorname{var}(\widehat{\mu}_{\mathrm{LB}}).$$

Moreover,

$$\begin{aligned}
\operatorname{var}(\widehat{\mu}_{\mathrm{LB}}) &= \operatorname{var}(w\, \overline{X}) \\
&= w^2 \left[E[\operatorname{var}(\overline{X} \mid \theta)] + \operatorname{var}(E(\overline{X} \mid \theta))\right] \\
&= w^2 \left[n^{-1} E[\operatorname{var}(X_1 \mid \theta)] + \operatorname{var}(\mu(\theta))\right] \\
&= w^2 \left[n^{-1} \varphi + \lambda\right] \\
&= \lambda \frac{n\lambda}{\varphi + n\lambda}.
\end{aligned}$$

Now the risk is given by

$$\rho(\widehat{\mu}_{\mathrm{LB}}) = \lambda - \lambda \frac{n\lambda}{\varphi + n\lambda} = (1 - w)\, \lambda$$

This concludes the proof. $\square$

In what follows, we suppress the dependence on the policy index i in the notation.

Example 6.3.2 (The linear Bayes estimator for Poisson distributed claim numbers and a gamma distributed heterogeneity parameter)

We assume the conditions of Example 5.2.4 and use the same notation. We want to calculate the linear Bayes estimator $\hat{\mu}_{\mathrm{LB}}$ for $\mu(\theta) = E(X_1|\theta) = \theta$. With $EX_1 = E\theta = \gamma/\beta$ and $\mathrm{var}(\theta) = \gamma/\beta^2$ we have

$$\varphi = E[\mathrm{var}(X_1 \mid \theta)] = E\theta = \gamma/\beta,$$

$$\lambda = \mathrm{var}(\theta) = \gamma/\beta^2.$$

Hence the weight w in (6.3.14) turns into

$$w = \frac{n\lambda}{\varphi + n\lambda} = \frac{\gamma/\beta^2}{\gamma/\beta + n\gamma/\beta^2} = \frac{n}{\beta + n}.$$

From Example 5.2.4 we conclude that the linear Bayes and the Bayes estimator coincide and have the same risk. In general we do not know the form of the Bayes estimator $\hat{\mu}_{\mathrm{B}}$ of $\mu(\theta)$ and therefore we cannot compare it with the linear Bayes estimator $\hat{\mu}_{\mathrm{LB}}$. □

Bühlmann [19] coined the name (linear) *credibility estimator* for the linear Bayes estimator

$$\hat{\mu}_{\mathrm{LB}} = (1 - w)\mu + w\overline{X}, \quad w = \frac{n\lambda}{\varphi + n\lambda} = \frac{n}{\varphi/\lambda + n},$$

w being the *credibility weight*. The larger w the more credible is the information contained in the data of the ith policy and the less important is the overall information about the portfolio represented by the expectation $\mu = E\mu(\theta)$. Since $w \to 1$ as $n \to \infty$ the credibility of the information in the policy increases with the sample size. But the size of w is also influenced by the ratio

$$\frac{\varphi}{\lambda} = \frac{E[\mathrm{var}(X_t \mid \theta)]}{\mathrm{var}(\mu(\theta))} = \frac{E[(X_t - \mu(\theta))^2]}{E[(\mu(\theta) - \mu)^2]}.$$

If φ/λ is small, w is close to 1. This phenomenon occurs if the variation of the claim sizes/claim numbers X_t in the individual policy is small compared to the variation in the whole portfolio. This can happen if there is a lot of heterogeneity in the portfolio, i.e., there is a lot of variation across the policies. This means that the expected claim size/claim number of the overall portfolio is quite meaningless when one has to determine the premium in a policy.

Any claim in the policy can be decomposed as follows

$$X_t = [X_t - \mu(\theta)] + [\mu(\theta) - \mu] + \mu. \tag{6.3.18}$$

The random variables $X_t - \mu(\theta)$ and $\mu(\theta) - \mu$ are uncorrelated. The quantity μ represents the expected claim number/claim size X_t in the portfolio. The difference $\mu(\theta) - \mu$ describes the deviation of the average claim number/claim size in the individual policy from the overall mean, whereas $X_t - \mu(\theta)$ is the (annual, say) fluctuation of the claim sizes/claim numbers X_t around the policy average. The credibility estimator $\widehat{\mu}_{LB}$ is based on the decomposition (6.3.18). The resulting formula for $\widehat{\mu}_{LB}$ as a weighted average of the policy and portfolio experience is essentially a consequence of (6.3.18).

Comments

Linear Bayes estimation seems to be quite restrictive since the random variable $\mu(\theta_i) = E(X_{i,t} \mid \theta_i)$ is approximated only by linear functions of the data $X_{i,t}$ in the ith policy. However, the general linear Bayes estimation procedure of Section 6.1 also allows one to calculate the minimum risk estimator of $\mu(\theta_i)$ in the class of all linear functions of any functions of the $X_{i,t}$'s which have finite variance. For example, the space $\mathcal{L}'$ introduced in (6.1.2) can be interpreted as the set of all linear functions of the powers $X_{i,t}^k$, $k \le p$, for some integer $p \ge 1$. Then minimum linear risk estimation amounts to the best approximation of $\mu(\theta_i)$ by all polynomials of the $X_{i,t}$'s of order p. We refer to Exercise 1 on p. 215 for an example with quadratic polynomials.

6.4 The Bühlmann-Straub Model

The Bühlmann model was further refined by Hans Bühlmann and Erwin Straub [20]. Their basic idea was to allow for heterogeneity inside each policy: each claim number/claim size $X_{i,t}$ is subject to an individual risk exposure expressed by an additional parameter $p_{i,t}$. These weights express our knowledge about the *volume* of $X_{i,t}$. For example, you may want to think of $p_{i,t}$ as the size of a particular house which is insured against fire damage or of the type of a particular car. In this sense, $p_{i,t}$ can be interpreted as *risk unit per time unit,* for example, per year.

In his monograph [75], Straub illustrated the meaning of *volume* by giving the different positions of the Swiss Motor Liability Tariff. The main positions are private cars, automobiles for goods transport, motor cycles, buses, special risks and short term risks. Each if these risks is again subdivided into distinct subclasses. He also refers to the positions of the German Fire Tariff which includes warehouses, mines and foundries, stone and earth, iron and metal works, chemicals, textiles, leather, paper and printing, wood, nutritionals, drinks and tobacco, and other risks. The variety of risks in these portfolios is rather high, and the notion of volume aims at assigning a quantitative measure for them.

Definition 6.4.1 (The Bühlmann-Straub model)
The model is defined by the requirements (1)-(3) in Definition 6.2.1, and Condition (4) is replaced by

(4') *Conditionally on θ_i, the $X_{i,t}$'s are independent and their expectation and variance are given functions of θ_i:*

$$\mu(\theta_i) = E(X_{i,t} \mid \theta_i) \quad and \quad \mathrm{var}(X_{i,t} \mid \theta_i) = v(\theta_i)/p_{i,t}\,.$$

The weights $p_{i,t}$ are pre-specified deterministic positive risk units.

Since the heterogeneity parameters θ_i are iid, the sequences $(\mu(\theta_i))$ and $(v(\theta_i))$ are iid.

We use the same notation as in the Bühlmann model

$$\mu = E\mu(\theta_i)\,, \quad \lambda = \mathrm{var}(\mu(\theta_i)) \quad and \quad \varphi = Ev(\theta_i)\,.$$

The following result is the analog of Theorem 6.3.1 for the linear Bayes estimator in the Bühlmann-Straub model.

Theorem 6.4.2 (Linear Bayes estimation in the Bühlmann-Straub model) *Assume $\mathrm{var}(X_{i,t}) < \infty$ for $i, t \geq 1$ and $\Sigma_{\mathbf{X}_i}$ is invertible for every i. Then the linear Bayes estimator $\widehat{\mu}_{\mathrm{LB}}$ of $\mu(\theta_i)$ in the class $\mathcal{L}$ of linear functions of the data $\mathbf{X}_1, \ldots, \mathbf{X}_r$ exists, is unique and given by*

$$\widehat{\mu}_{\mathrm{LB}} = (1 - w)\,\mu + w\,\overline{X}_{i,\cdot}\,,$$

where

$$w = \frac{\lambda\,p_{i,\cdot}}{\varphi + \lambda\,p_{i,\cdot}} \quad and \quad \overline{X}_{i,\cdot} = \frac{1}{p_{i,\cdot}} \sum_{t=1}^{n_i} p_{i,t}\,X_{i,t}$$

The risk of $\widehat{\mu}_{\mathrm{LB}}$ is given by

$$\rho(\widehat{\mu}_{\mathrm{LB}}) = (1 - w)\,\lambda\,.$$

The proof of this result is completely analogous to the Bühlmann model (Theorem 6.3.1) and left as an exercise. We only mention that the normal equations in the ith portfolio, see Proposition 6.1.1, and the corresponding relations (6.3.16) and (6.3.17) in the proof of Theorem 6.3.1 boil down to the equations

$$a_0 = \mu\,(1 - a_{i,\cdot})\,,$$

$$\lambda = \lambda\,a_{i,\cdot} + \varphi\,\frac{a_{i,t}}{p_{i,t}}\,, \quad t = 1, \ldots, n\,.$$

Comments

In the Bühlmann and Bühlmann-Straub models the global parameters μ, φ, λ of the portfolio have to be estimated from the data contained in all policies. In the exercises below we hint at some possible estimators of these quantities; see also the references below.

The classical work on credibility theory and experience rating is summarized in Bühlmann's classic text [19]. A more recent textbook treatment aimed at actuarial students is Kaas et al. [46]. Textbook treatments of credibility theory and related statistical questions can be found in the textbooks by Klugman et al. [51], Sundt [77], Straub [75].

Exercises

(1) We consider the ith policy in the heterogeneity model and suppress the dependence on i in the notation. Assume we have one claim number X in the policy which is $\text{Pois}(\theta)$ distributed, given some positive random variable θ. Assume that the moments $m_k = E(\theta^k) < \infty$, $k = 1, 2, 3, 4$, are known.

(a) Determine the linear Bayes estimator $\widehat{\theta}$ for $\mu(\theta) = E(X \mid \theta) = \theta$ based on X only in terms of X, m_1, m_2. Express the minimal linear Bayes risk $\rho(\widehat{\theta})$ as a function of m_1 and m_2.

(b) Now we want to find the best estimator $\widetilde{\theta}_{\text{LB}}$ of θ with respect to the quadratic risk $\rho(\widetilde{\mu}) = E[(\theta - \widetilde{\theta})^2]$ in the class of linear functions of X and $X(X-1)$:

$$\widetilde{\theta} = a_0 + a_1 X + a_2 X(X-1), \qquad a_0, a_1, a_2 \in \mathbb{R}.$$

This means that $\widetilde{\theta}$ is the linear Bayes estimator of θ based on the data $\mathbf{X} = (X, X(X-1))'$. Apply the normal equations to determine a_0, a_1, a_2. Express the relevant quantities by the moments m_k.

Hint: Use the well-known identity $EY^{(k)} = \lambda^k$ for the factorial moments $EY^{(k)} = E[Y(Y-1)\cdots(Y-k+1)]$, $k \geq 1$, of a random variable $Y \sim \text{Pois}(\lambda)$.

(2) For Exercise 2 on p. 200 calculate the linear Bayes estimate of $p(\theta) = E(Y_1 \mid \theta)$ based on the data $Y_1, \ldots, Y_n$ and the corresponding linear Bayes risk. Compare the Bayes and the linear Bayes estimators and their risks.

(3) For Exercise 4 on p. 201 calculate the linear Bayes estimator of $E(X_1 \mid \theta)$ and the corresponding linear Bayes risk. Compare the Bayes and the linear Bayes estimators and their risks.

(4) For Exercise 5 on p. 201 calculate the linear Bayes estimator of $E(X_1 \mid \theta)$ and the corresponding linear Bayes risk. Compare the Bayes and the linear Bayes estimators and their risks.

(5) Consider a portfolio with n independent policies.

(a) Assume that the claim numbers $X_{i,t}$, $t = 1, 2, \ldots$, in the ith policy are independent and $\text{Pois}(p_{i,t}\theta_i)$ distributed, given θ_i. Assume that $p_{i,t} \neq p_{i,s}$ for some $s \neq t$. Are the conditions of the Bühlmann-Straub model satisfied?

(b) Assume that the claim sizes $X_{i,t}$, $t = 1, 2, \ldots$, in the ith policy are independent and $\Gamma(\gamma_{i,t}, \beta_{i,t})$ distributed, given θ_i. Give conditions on $\gamma_{i,t}$, $\beta_{i,t}$ under which the Bühlmann-Straub model is applicable. Identify the parameters μ, φ, λ and $p_{i,t}$.

(6) Consider the Bühlmann-Straub model with r policies, where the claim sizes/claim numbers $X_{i,t}$, $t = 1, 2, \ldots$, in policy i are independent, given θ_i. Let w_i be positive weights satisfying $\sum_{i=1}^n w_i = 1$ and $\overline{X}_{i,\cdot} = p_{i\cdot}^{-1} \sum_{t=1}^{n_i} p_{i,t} X_{i,t}$ be the (weighted) sample mean in the ith policy.

(a) Show that

$$\widehat{\mu} = \sum_{i=1}^r w_i \overline{X}_{i,\cdot}. \tag{6.4.19}$$

is an unbiased estimator of $\mu = E\mu(\theta_i) = E(X_{i,t} \mid \theta_i)$.

(b) Calculate the variance of $\widehat{\mu}$ in (6.4.19).

(c) Choose the weights w_i in such a way that $\text{var}(\widehat{\mu})$ is minimized and calculate the minimal value $\text{var}(\widehat{\mu})$.

(7) Consider the Bühlmann-Straub model.

(a) Guess what is estimated by the statistics

$$s_1 = \sum_{i=1}^{n} \sum_{t=1}^{n} p_{i,t} \left(X_{i,t} - \overline{X}_{i \cdot} \right)^2 \quad \text{and} \quad s_2 = \sum_{i=1}^{n} w_i \left(\overline{X}_{i \cdot} - \widehat{\mu} \right)^2 ,$$

where w_i are the optimal weights derived in Exercise 6 above and $\widehat{\mu}$ is defined in (6.4.19).

(b) Calculate the expectations of s_1 and s_2. Are your guesses from (a) confirmed by these calculations?

(c) Calculate Es_2 with the weights $w_i = p_{i \cdot}/p_{\cdot \cdot}$, where $p_{\cdot \cdot} = \sum_{i=1}^{n} p_{i \cdot}$. Modify s_1 and s_2 such they become unbiased estimators of the quantities which are suggested by (b).

References

Each reference is followed, in square brackets, by a list of the page numbers where this reference is cited.

1. ALSMEYER, G. (1991) *Erneuerungstheorie*. Teubner, Stuttgart. *[71]*
2. ANDERSEN, P.K., BORGAN, Ø., GILL, R.D. AND KEIDING, N. (1993) *Statistical Models Based on Counting Processes*. Springer, New York. *[9]*
3. ASMUSSEN, S. (1999) *Stochastic Simulation With a View Towards Stochastic Processes*. MaPhySto Lecture Notes. Available under www.maphysto.dk. *[137,143]*
4. ASMUSSEN, S. (2000) *Ruin Probabilities*. World Scientific, Singapore. *[130,137, 143,166,181]*
5. ASMUSSEN, S. (2003) *Applied Probability and Queues*. Springer, Berlin. *[71]*
6. ASMUSSEN, S., BINSWANGER, K. AND HØJGAARD, B. (2000) Rare event simulation for heavy-tailed distributions. *Bernoulli* **6**, 303–322. *[143]*
7. ASMUSSEN, S. AND RUBINSTEIN, R.Y. (1995) Steady-state rare events simulation in queueing models and its complexity properties. In: Dshalalow, J. (Ed.) *Advances in Queueing: Models, Methods and Problems*, pp. 429–466. CRC Press, Boca Raton. *[143]*
8. BARBOUR, A.D., HOLST, L. AND JANSON, S. (1992) *Poisson Approximation*. Oxford University Press, New York. *[138]*
9. BARNDORFF-NIELSEN, O.E., MIKOSCH, T. AND RESNICK, S.I. (Eds.) (2002) *Lévy Processes: Theory and Applications*. Birkhäuser, Boston. *[18]*
10. BEIRLANT, J., TEUGELS, J.L. AND VYNCKIER, P. (1996) *Practical Analysis of Extreme Values*. Leuven University Press, Leuven. *[97, 99]*
11. BICKEL, P. AND FREEDMAN, D. (1981) Some asymptotic theory for the bootstrap. *Ann. Statist.* **9**, 1196–1217. *[141,146]*
12. BILLINGSLEY, P. (1968) *Convergence of Probability Measures*. Wiley, New York. *[15,146,161]*
13. BILLINGSLEY, P. (1995) *Probability and Measure*. 3rd edition. Wiley, New York. *[14,27,30,45,82,90,138,153,174]*

218   References

14. BINGHAM, N.H., GOLDIE, C.M. AND TEUGELS, J.L. (1987) *Regular Variation.* Cambridge University Press, Cambridge. *[106,109,177]*

15. BJÖRK, T. (1999) *Arbitrage Theory in Continuous Time.* Oxford University Press, Oxford (UK). *[148]*

16. BROCKWELL, P.J. AND DAVIS, R.A. (1991) *Time Series: Theory and Methods.* 2nd edition. Springer, New York. *[39,174,199]*

17. BÜHLMANN, H. (1967) Experience rating and credibility I. *ASTIN Bulletin* **4**, 199–207. *[189,199]*

18. BÜHLMANN, H. (1969) Experience rating and credibility II. *ASTIN Bulletin* **5**, 157–165. *[189]*

19. BÜHLMANN, H. (1970) *Mathematical Methods in Risk Theory.* Springer, Berlin. *[86,189,212,214]*

20. BÜHLMANN, H. AND STRAUB, E. (1970) Glaubwürdigkeit für Schadensätze. *Mittl. Ver. Schw. Vers. Math.* **73**, 205–216. *[213]*

21. CHAMBERS, J.M. (1977) *Computational Methods for Data Analysis.* Wiley, New York. Wadsworth, Belmont Ca., Duxbury Press, Boston. *[90]*

22. CHOSSY, R. VON AND RAPPL, G. (1983) Some approximation methods for the distribution of random sums. *Ins. Math. Econ.* **2**, 251–270. *[132]*

23. CRAMÉR, H. (1930) *On the mathematical theory of risk.* Skandia Jubilee Volume, Stockholm. Reprinted in: Martin-Löf, A. (Ed.) Cramér, H. (1994) *Collected Works.* Springer, Berlin. *[155,166]*

24. CRESSIE, N.A.C. (1993) *Statistics for Spatial Data.* Wiley, New York. *[48,199]*

25. DALEY, D. AND VERE-JONES, D. (1988) *An Introduction to the Theory of Point Processes.* Springer, Berlin. *[52]*

26. EFRON, B. (1979) Bootstrap methods: another look at the jackknife. *Ann. Statist.* **7**, 1–26. *[138,141]*

27. EFRON, B. AND TIBSHIRANI, R.J. (1993) *An Introduction to the Bootstrap.* Chapman and Hall, New York. *[141,143]*

28. EMBRECHTS, P., GRÜBEL, R. AND PITTS, S.M. (1993) Some applications of the fast Fourier transform algorithm in insurance mathematics. *Statist. Neerlandica* **47**, 59–75. *[130]*

29. EMBRECHTS, P., KLÜPPELBERG, C. AND MIKOSCH, T. (1997) *Modelling Extremal Events for Insurance and Finance.* Springer, Heidelberg. *[52,65,82,102, 104,105,107,109–112,135,141,142,149,154,171,175,180]*

30. EMBRECHTS, P. AND VERAVERBEKE, N. (1982) Estimates for the probability of ruin with special emphasis on the possibility of large claims. *Insurance: Math. Econom.* **1**, 55–72. *[178]*

31. FAN, J. AND GIJBELS, I. (1996) *Local Polynomial Modelling and Its Applications.* Chapman & Hall, London. *[39]*

32. FELLER, W. (1971) *An Introduction to Probability Theory and Its Applications II.* Wiley, New York. *[56,71,109,177,183,185]*

33. GASSER, T., ENGEL, J. AND SEIFERT, B. (1993) Nonparametric function estimation. In: *Computational Statistics. Handbook of Statistics.* Vol. **9**, pp. 423–465. North-Holland, Amsterdam. *[39]*

34. GOLDIE, C.M. (1991) Implicit renewal theory and tails of solutions of random equations. *Ann. Appl. Probab.* **1**, 126–166. *[175]*

35. GOLDIE, C.M. AND KLÜPPELBERG, C. (1996) Subexponential distributions. In: Adler, R., Feldman, R. and Taqqu, M.S. (Eds.) *A Practical Guide to Heavy Tails: Statistical Techniques for Analysing Heavy-Tailed Distributions*, pp. 435–460. Birkhäuser, Boston. *[111]*

36. GRANDELL, J. (1991) *Aspects of Risk Theory.* Springer, Berlin. *[180]*

37. GRANDELL, J. (1997) *Mixed Poisson Processes.* Chapman and Hall, London. *[74]*

38. GRÜBEL, R. AND HERMESMEIER, R. (1999) Computation of compound distributions I: aliasing erros and exponential tilting. *ASTIN Bulletin* **29**, 197–214. *[130]*

39. GRÜBEL, R. AND HERMESMEIER, R. (2000) Computation of compound distributions II: discretization errors and Richardson extraploation. *ASTIN Bulletin* **30**, 309–331. *[130]*

40. GUT, A. (1988) *Stopped Random Walks.* Springer, Berlin. *[65,71]*

41. HALL, P.G. (1992) *The Bootstrap and Edgeworth Expansion.* Springer, New York. *[132,143]*

42. HEIDELBERGER, P. (1995) Fast simulation of rare events in queueing and reliability models. *ACM TOMACS* **6**, 43–85. *[143]*

43. HESS, K.T., LIEWALD, A. AND SCHMIDT, K.D. (2002) An extension of Panjer's recursion. *ASTIN Bulletin* **32**, 283–298. *[130]*

44. HOGG, R.V. AND KLUGMAN, S.A. (1984) *Loss Distributions.* Wiley, New York. *[99]*

45. JENSEN, J. (1995) *Saddlepoint Approximations.* Oxford University Press, Oxford. *[132]*

46. KAAS, R., GOOVAERTS, M., DHAENE, J. AND DENUIT, M. (2001) *Modern Actuarial Risk Theory.* Kluwer, Boston. *[86,88,130,199,214]*

47. KALLENBERG, O. (1973) Characterization and convergence of random measures and point processes. *Z. Wahrscheinlichkeitstheorie verw. Geb.* **27**, 9–21. *[74]*

48. KALLENBERG, O. (1983) *Random Measures.* 3rd edition. Akademie-Verlag, Berlin. *[52]*

49. KESTEN, H. (1973) Random difference equations and renewal theory for products of random matrices. *Acta Math.* **131**, 207–248. *[175]*

50. KINGMAN, J.F.C. (1996) *Poisson Processes.* Oxford University Press, Oxford (UK). *[52]*

51. KLUGMAN, S.A., PANJER, H.H. AND WILLMOT, G.E. (1998) *Loss Models. From Data to Decisions.* Wiley, New York. *[86,199,214]*

52. KOLLER, M. (2000) *Stochastische Modelle in der Lebensversicherung.* Springer, Berlin. *[19]*

53. LEHMANN, E.L. (1986) *Testing Statistical Hypotheses.* Springer, New York. *[199]*

54. LUKACS, E. (1970) *Characteristic Functions.* 2nd edition. Hafner Publ. Co., New York. *[53,145]*

55. LUNDBERG, F. (1903) *Approximerad framställning av sannolikhetsfunktionen. Återförsäkring av kollektivrisker.* Akad. Afhandling. Almqvist och Wiksell, Uppsala. *[7,9]*

56. MAMMEN, E. (1992) *When Does Bootstrap Work?* Asymptotic Results and Simulations. *Lecture Notes in Statistics* **77**, Springer, New York. *[143]*

57. MIKOSCH, T. (1998) *Elementary Stochastic Calculus with Finance in View.* World Scientific, Singapore. *[18,148]*

58. MIKOSCH, T. (2003) Modelling dependence and tails of financial time series. In: Finkenstädt, B. and Rootzén, H. (Eds.) *Extreme Values in Finance, Telecommunications and the Environment.* CRC Press. To appear. *[48,175]*

59. MÜLLER, H.-G. AND STADTMÜLLER, U. (1987) Estimation of heteroscedasticity in regression analysis. *Ann. Statist.* **15**, 610–625. *[39]*

60. PANJER, H.H. (1981) Recursive evaluation of a family of compound distributions. *ASTIN Bulletin* **11**, 22–26. *[126]*

61. PETROV, V.V. (1975) *Sums of Independent Random Variables.* Springer, Berlin. *[132]*

62. PITMAN, E.J.G. (1980) Subexponential distribution functions. *J. Austral. Math. Soc. Ser. A* **29**, 337–347. *[114]*

63. PRIESTLEY, M.B. (1981) *Spectral Analysis and Time Series.* Vols. **I** and **II**. Academic Press, New York. *[39]*

64. RESNICK, S.I. (1987) *Extreme Values, Regular Variation, and Point Processes.* Springer, New York. *[48,52,58,90,109]*

65. RESNICK, S.I. (1992) *Adventures in Stochastic Processes.* Birkhäuser, Boston. *[19,23,52,63,67–71,161,177]*

66. ROGERS, L.C.G. AND WILLIAMS, D. (2000) *Diffusions, Markov Processes and Martingales.* Vol. **1**. Cambridge University Press, Cambridge (UK). *[14,18,19]*

67. ROLSKI, T., SCHMIDLI, H., SCHMIDT, V. AND TEUGELS, J. (1999) *Stochastic Processes for Insurance and Finance.* Wiley, New York. *[71,130,166,180,181]*

68. ROOTZÉN, H. AND TAJVIDI, N. (1997) Extreme value statistics and wind storm losses: a case study. *Scand. Actuar. J.* 70–94. *[97]*

69. RYTGAARD, M. (1996) Simulation experiments on the mean residual life function $m(x)$. In: *Proceedings of the XXVII ASTIN Colloquium, Copenhagen, Denmark.* Vol. **1**, pp. 59–81. *[97]*

70. SAMORODNITSKY, G. AND TAQQU, M.S. (1994) *Stable Non-Gaussian Random Processes. Stochastic Models with Infinite Variance.* Chapman and Hall, London. *[57]*

71. SATO, K.-I. (1999) *Lévy Processes and Infinitely Divisible Distributions.* Cambridge University Press, Cambridge (UK). *[15,18,145]*

72. SCHROEDER, M. (1990) *Fractals, Chaos, Power Laws.* Freeman, New York. *[106]*

73. SIGMA (2003) Tables on the major losses 1970-2002. *Sigma publication No 2,* Swiss Re, Zürich, p. 34. The publication is available under www.swissre.com. *[102]*

74. SPITZER, F. (1976) *Principles of Random Walk.* 2nd edition. Springer, Berlin. *[159]*

75. STRAUB, E. (1988) *Non-Life Insurance Mathematics.* Springer, New York. *[199,213,214]*

76. SUNDT, B. (1999) *An Introduction to Non-Life Insurance Mathematics.* 4th edition. VVW Karlsruhe. *[199,214]*

77. SUNDT, B. (1999) On multivariate Panjer recursions. *ASTIN Bulletin* **29**, 29–46. *[130]*

78. WILLIAMS, D. (1991) *Probability with Martingales*. Cambridge University Press, Cambridge (UK). *[62,65,65,135,197]*

79. WILLINGER, W., TAQQU, M.S., LELAND, M. AND WILSON. D. (1995) Self-similarity through high variability: statistical analysis of ethernet lan traffic at the source level. In: *Proceedings of the ACM/SIGCOMM'95, Cambridge, MA. Computer Communications Review* **25**, 100–113. *[66]*

80. WILLMOT, G.E. AND LIN, X.S. (2001) *Lundberg Approximations for Compound Distributions with Insurance Applications*. Springer, Berlin. *[130]*

Index

S

List of Abbreviations and Symbols

We have tried as much as possible to use uniquely defined abbreviations and symbols. In various cases, however, symbols can have different meanings in different sections. The list below gives the most typical usage. Commonly-used mathematical symbols are not explained here.

Abbreviation or Symbol	Explanation	p.
a.s.	almost sure, almost surely, with probability 1	
a.e.	almost everywhere, almost every	
$\text{Bin}(n, p)$	binomial distribution with parameters (n, p): $p(k) = \binom{n}{k} p^k (1 - p)^{n-k}$, $k = 0, \ldots, n$	
$\mathbb{C}$	set of the complex numbers	
$\text{corr}(X, Y)$	correlation between the random variables X and Y	
$\text{cov}(X, Y)$	covariance between the random variables X and Y	
$E_F X$	expectation of X with respect to the distribution F	
$e_F(u)$	mean excess function	94
$\text{Exp}(\lambda)$	exponential distribution with parameter λ: $F(x) = 1 - \mathrm{e}^{-\lambda x}$, $x > 0$	
F	distribution function/distribution of a random variable	
F_A	distribution function/distribution of the random variable A	
F_I	integrated tail distribution: $F_I(x) = (E_F X)^{-1} \int_0^x \overline{F}(y)\,dy$, $x \geq 0$	167
F_n	empirical (sample) distribution function	89
$F^{\leftarrow}(p)$	p-quantile/quantile function of F	89
$F_n^{\leftarrow}(p)$	empirical p-quantile	90
$\overline{F}$	tail of the distribution function F: $\overline{F} = 1 - F$	
F^{n*}	n-fold convolution of the distribution function/distribution F	
$\widehat{f}_X$	Laplace-Stieltjes transform of the random variable X:	

	$\widehat{f}_X(s) = E e^{-sX}$, $s > 0$	177
Γ	gamma function : $\Gamma(x) = \int_0^\infty t^{x-1} e^{-t} dt$	
$\Gamma(\gamma, \beta)$	gamma distribution with parameters γ and β:	
	gamma density $f(x) = \beta^\gamma (\Gamma(\gamma))^{-1} x^{\gamma-1} e^{-\beta x}$, $x > 0$	
IBNR	incurred but not reported claim	48
I_A	indicator function of the set (event) A	
iid	independent, identically distributed	
λ	intensity or intensity function of a Poisson process	15
Λ	Gumbel distribution: $\Lambda(x) = \exp\{-e^{-x}\}$, $x \in \mathbb{R}$	154
Leb	Lebesgue measure	
$\log x$	logarithm with basis e	
$\log^+ x$	$\log^+ x = \max(\log x, 0)$	
$L(x)$	slowly varying function	105
M_n	maximum of $X_1, \ldots, X_n$	
$\mu(t)$	mean value function of a Poisson process on $[0, \infty)$	14
$\mathbb{N}$	set of the positive integers	
$\mathbb{N}_0$	set of the non-negative integers	
$N, N(t)$	claim number or claim number process	7
$\widetilde{N}$	often a homogeneous Poisson process	
$N(\mu, \sigma^2)$	Gaussian (normal) distribution with mean μ, variance σ^2	
$N(0, 1)$	standard normal distribution	
$N(\mu, \Sigma)$	multivariate Gaussian (normal) distribution with mean	
	vector μ and covariance matrix Σ	
NPC	net profit condition	159
$o(1)$	$h(x) = o(1)$ as $x \to x_0 \in [-\infty, \infty]$ means that	
	$\lim_{x \to x_0} h(x) = 0$	20
ω	$\omega \in \Omega$ random outcome	
$(\Omega, \mathcal{F}, P)$	probability space	
$\phi_X(t)$	characteristic function of the random variable X:	
	$\phi_X(t) = E e^{itX}$, $t \in \mathbb{R}$	
Φ	standard normal distribution/distribution function	
Φ_α	Frechet distribution: $\Phi_\alpha(x) = \exp\{-x^{-\alpha}\}$, $x > 0$	154
$\text{Pois}(\lambda)$	Poisson distribution with parameter λ:	
	$p(n) = e^{-\lambda} \lambda^n / n!$, $n \in \mathbb{N}_0$	
PRM	Poisson random measure	
$\text{PRM}(\mu)$	Poisson random measure with mean measure μ	46
$\psi(u)$	ruin probability	157
Ψ_α	Weibull (extreme value) distribution:	
	$\Psi_\alpha(x) = \exp\{-(-x)^\alpha\}$, $x < 0$	154
$\mathbb{R}, \mathbb{R}^1$	real line	
$\mathbb{R}_+$	$\mathbb{R}_+ = (0, \infty)$	
$\mathbb{R}^d$	d-dimensional Euclidean space	
ρ	safety loading	85
$\rho(\widehat{\mu})$	(quadratic) Bayes or linear Bayes risk of $\widehat{\mu}$	194

$\mathcal{S}$	class of the subexponential distributions	109
$\mathrm{sign}(a)$	sign of the real number a	
S_n	cumulative sum of $X_1, \ldots, X_n$	
$S, S(t)$	total, aggregate claim amount process	8
t	time, index of a stochastic process	
T_i	arrival times of a claim number process	7
u	initial capital	156
$\mathrm{U}(a, b)$	uniform distribution on (a, b)	
$U(t)$	risk process	156
$\mathrm{var}(X)$	variance of the random variable X	
$\mathrm{var}_F(X)$	variance of a random variable X with distribution F	
X_n	claim size	7
$X_{(n-i+1)}$	ith largest order statistic in the sample $X_1, \ldots, X_n$	28
$\overline{X}_n$	sample mean	
$\mathbb{Z}$	set of the integers	
$\sim$	$X \sim F$: X has distribution F	
$\approx$	$a(x) \approx b(x)$ as $x \to x_0$ means that $a(x)$ is approximately (roughly) of the same order as $b(x)$ as $x \to x_0$. It is only used in a heuristic sense.	
$*$	convolution or bootstrapped quantity	
$\|\cdot\|$	$\|x\|$ norm of x	
$[\cdot]$	$[x]$ integer part of x	
$\{\cdot\}$	$\{x\}$ fractional part of x	
x_+	positive part of a number: $x_+ = \max(0, x)$	
B^c	complement of the set B	
$\xrightarrow{\mathrm{a.s.}}$	$A_n \xrightarrow{\mathrm{a.s.}} A$: a.s. convergence	
$\xrightarrow{d}$	$A_n \xrightarrow{d} A$: convergence in distribution	
$\xrightarrow{P}$	$A_n \xrightarrow{P} A$: convergence in probability	
$\stackrel{d}{=}$	$A \stackrel{d}{=} B$: A and B have the same distribution	

For a function f on $\mathbb{R}$ and intervals $(a, b]$, $a < b$, we write $f(a, b] = f(b) - f(a)$.

Universitext

Druck und Bindung: Strauss Offsetdruck GmbH